Life
on the
Edge

www.**transworldbooks**.co.uk

—

Also by Jim Al-Khalili:

Black Holes, Wormholes and Time Machines
Nucleus: A Trip into the Heart of Matter
Paradox: The Nine Greatest Enigmas in Science

Also by Johnjoe McFadden:

Quantum Evolution
Human Nature: Fact and Fiction

Life
on the
Edge

The Coming of Age
of Quantum Biology

Jim Al-Khalili
and
Johnjoe McFadden

BANTAM PRESS

LONDON • TORONTO • SYDNEY • AUCKLAND • JOHANNESBURG

TRANSWORLD PUBLISHERS
61–63 Uxbridge Road, London W5 5SA
A Random House Group Company
www.transworldbooks.co.uk

First published in Great Britain
in 2014 by Bantam Press
an imprint of Transworld Publishers

A CIP catalogue record for this book
is available from the British Library.

ISBNs 9780593069318 (cased)
9780593069325 (tpb)

Line illustrations by HL Studios

Addresses for Random House Group Ltd companies outside the UK
can be found at: www.randomhouse.co.uk
The Random House Group Ltd Reg. No. 954009

The Random House Group Limited supports the Forest Stewardship Council® (FSC®),
the leading international forest-certification organisation. Our books carrying the
FSC label are printed on FSC®-certified paper. FSC is the only forest-certification
scheme supported by the leading environmental organisations, including
Greenpeace. Our paper procurement policy can be found at
www.randomhouse.co.uk/environment

Typeset in 11.5/15.5 pt Minion by
Jouve (UK), Milton Keynes
Printed and bound in Great Britain by
Clays Ltd, Bungay, Suffolk

2 4 6 8 10 9 7 5 3 1

For
Penny and Ollie
Julie, David and Kate

Contents

Acknowledgements

THIS BOOK has been three years in the writing, although the authors have collaborated on research in this exciting new field, which brings quantum physics, biochemistry and biology together, for almost two decades. But when it comes to such a cross-disciplinary area of science as quantum biology, it is impossible ever to become expert enough to explain in sufficient depth and with sufficient confidence all the science that is needed to paint the full picture – particularly when it comes to writing the first ever book on the subject for a lay audience.

It is certainly true that neither of the authors could have written this book alone, since we each bring our own expertise from the worlds of physics and biology, respectively, to the table. It is even truer that we would not have been able to produce a book that we are both immensely proud of without the help and advice of many people, most of whom are world leaders in their areas of research.

We are grateful to Paul Davies for many fruitful discussions he has had with both of us over the past fifteen years about quantum mechanics and its potential relevance in biology. We are also indebted to the many physicists, chemists and biologists now making great strides in this new field whose expertise and deep knowledge of their specialist areas we did not, and do not, have. In particular, we are indebted to Jennifer Brookes, Gregory Engel, Adam Godbeer, Seth Lloyd, Alexandra Olaya-Castro, Martin Plenio, Sandu Popescu, Thorsten Ritz, Gregory Scholes, Nigel Scrutton, Paul Stevenson,

Luca Turin and Vlatko Vedral. We also wish to thank Mirela Dumic, the coordinator of the University of Surrey's Institute of Advanced Studies, who almost single-handedly put together our highly success-ful international workshop, 'Quantum Biology: Current Status and Opportunities', at Surrey in 2012, which was jointly funded by the IAS, BBSRC (Biotechnology and Biological Sciences Research Council) and MILES (Models and Mathematics in Life and Social Sciences) project. This workshop brought together many of the leading figures – this is still an emerging field and the number working in it is relatively small – currently involved in quantum biology research from around the world and helped us feel as though we were truly part of this exciting research community.

Once the book was in draft form, we asked several of those col-leagues listed above to read through it and give us their opinions. We are thus especially grateful to Martin Plenio, Jennifer Brookes, Alexandra Olaya-Castro, Gregory Scholes, Nigel Scrutton and Luca Turin. We would also like to thank Philip Ball, Pete Downes and Greg Knowles for reading through some or all of the final draft and pro-viding so many insightful and useful comments that have improved the book tremendously. A big thank-you goes to our agent, Patrick Walsh, without whom the book would not have got off the ground, and to Sally Gaminara at Random House, for her faith in us and for being so excited about the project. An even bigger thank-you has to go to Patrick and to Carrie Plitt at Conville & Walsh, for their advice and suggestions about the book's structure and format, and in help-ing to mould it into a final version that is light years away from its initial clunky state. We are also indebted to Gillian Somerscales for her editorial brilliance.

Last, but by no means least, we wish to thank our families for their unstinting support, particularly during those periods when we were facing self- and publisher-imposed deadlines, which meant putting all other commitments to one side and shutting ourselves away with our laptops. We have lost count of the evenings, weekends and family vacations during which quantum biology has had to come first. We hope the book has been worth it.

For both for us, and for the new field of quantum biology, we hope that journey has only just begun.

Jim Al-Khalili and Johnjoe McFadden
August 2014

1
Introduction

THE WINTER frost has arrived early this year in Europe and there is a penetrating chill in the evening air. Buried deep within a young robin's mind, a once vague sense of purpose and resolve grows stronger.

The bird has spent the past few weeks devouring far more than her normal intake of insects, spiders, worms and berries and is now almost double the weight that she was when her brood flew the nest back in August. This extra bulk is mostly fat reserves, which she will require as fuel for the arduous journey upon which she is about to embark.

This will be her first migration away from the spruce forest in central Sweden where she has lived for the duration of her short life and where she reared her young chicks just a few months ago. Luckily for her, the previous winter was not too harsh, for a year ago she was not yet fully grown and therefore not strong enough to undertake such a long journey. But now, with her parental responsibilities discharged until next spring, she has only herself to think about, and she is ready to escape the coming winter by heading south to seek a warmer climate.

It is a couple of hours after sunset. Rather than settle for the night, she hops in the gathering gloom to the tip of a branch near the base of the huge tree that she has made her home since the spring. She gives herself a quick shake, much like a marathon runner loosening up her muscles before a race. Her orange breast glistens in the moonlight. The painstaking effort and care she invested in building her nest – just a few feet away, partially hidden against the moss-covered bark of the tree trunk – is now a dim memory.

She is not the only bird preparing to depart, for other robins – both male and female – have also decided that this is the right night to begin their long migration south. In the trees all around her she hears loud, shrill singing that drowns out the usual sounds of other nocturnal woodland creatures. It is as though the birds feel compelled to announce their departure, sending out a message to the other forest inhabitants that they should think twice before contemplating invading the birds' territory and empty nests while they are gone. For these robins most certainly plan to be back in the spring.

With a quick tilt of her head this way and that to make sure the coast is clear, she takes off into the evening sky. The nights have been lengthening with winter's advance and she will have a good ten hours or so of flying ahead of her before she can rest again.

She sets off on a course bearing of 195° (15° to the west of due south). Over the coming days she will carry on flying in, more or less, this same direction, covering two hundred miles on a good day. She has no idea what to expect along the journey, nor any sense of how long it will take. The terrain around her spruce wood is a familiar one, but after a few miles she is flying over an alien moonlit landscape of lakes, valleys and towns.

Somewhere near the Mediterranean she will arrive at her destination; although she is not heading for any specific location, when she does arrive at a favourable spot she will stop, memorizing the local landmarks so that she can return there in the coming years. If she has the strength, she may even fly all the way across to the North African coast. But this is her first migration, and her only priority now is to escape the biting cold of the approaching Nordic winter.

She seems oblivious to the surrounding robins that are all flying in roughly the same direction, some of which will have made the journey many times before. Her night vision is superb, but she is not looking for any landmarks – as we might were we making such a journey – nor is she tracking the pattern of the stars in the clear night sky by consulting her internal celestial map, as many other nocturnal migrating birds do. Instead, she has a rather remarkable skill and several million years of evolution to thank for her capacity to make what will become an annual autumn migration, a trip of some two thousand miles.

Migration is, of course, commonplace in the animal kingdom. Every winter, for instance, salmon spawn in the rivers and lakes of northern Europe, leaving young fry that, after hatching, follow the course of their river out to sea and into the North Atlantic, where they grow and mature; three years later, these young salmon return to breed in the same rivers and lakes where they spawned. New World monarch butterflies migrate thousands of miles southward across the entire United States in the autumn. They, or their descendants (as they will breed en route), then return north to the same trees in which they pupated in the spring. Green turtles that hatch on the shores of Ascension Island in the South Atlantic swim across thousands of miles of ocean before returning, every three years, to breed on the exact same eggshell-littered beach from which they emerged. The list goes on: many species of birds, whales, caribou, spiny lobsters, frogs, salamanders and even bees are all capable of undertaking journeys that would challenge the greatest human explorers.

How animals manage to find their way around the globe has been a mystery for centuries. We now know that they employ a variety of methods: some use solar navigation during the day and celestial navigation at night; some memorize landmarks; others can even *smell* their way around the planet. But the most mysterious navigational sense of all is the one possessed by the European robin: the ability to detect the direction and strength of the earth's magnetic field, known as magnetoreception. And while we now know of a number of other creatures that possess this ability, it is the way the European robin (*Erithacus rubecula*) finds her way across the globe that is of greatest interest to our story.

The mechanism that enables our robin to know how far to fly, and in which direction, is encoded in the DNA she inherited from her parents. This ability is a sophisticated and unusual one – a *sixth sense* that she uses to plot her course. For, like many other birds, and indeed insects and marine creatures, she has the ability to sense the earth's weak magnetic field and to draw directional information from it by way of an inbuilt navigational sense, which in her case requires a novel type of chemical compass.

Magnetoreception is an enigma. The problem is that the earth's magnetic field is very weak – between 30 and 70 microtesla at the surface: sufficient to deflect a finely balanced and almost frictionless compass needle, but only about a hundredth the force of a typical fridge magnet. This presents a puzzle: for the earth's magnetic field to be detected by an animal it must somehow influence a chemical reaction somewhere in the animal's body – this is, after all, how all living creatures, ourselves included, sense any external signal. But the amount of energy supplied by the interaction of the earth's magnetic field with the molecules within living cells is less than a billionth of the energy needed to break or make a chemical bond. How, then, can that magnetic field be perceptible to the robin?

Mysteries, however small, are fascinating because there's always the possibility that their solution may lead to a fundamental shift in our understanding of the world. Copernicus's ponderings in the sixteenth century on a relatively minor problem concerning the geometry of the Ptolemaic geocentric model of the solar system, for instance, led him to shift the centre of gravity of the entire universe away from humankind. Darwin's obsession with the geographical distribution of animal species and the mystery of why isolated island species of finches and mockingbirds tend to be so specialized led him to propose his theory of evolution. And German physicist Max Planck's solution to the mystery of blackbody radiation, concerning the way warm objects emit heat, led him to suggest that energy came in discrete lumps called 'quanta', leading to the birth of quantum theory in the year 1900. So, could the solution to the mystery of how birds find their way around the globe lead to a revolution in biology? The answer, bizarre as it may seem, is: yes.

But mysteries such as this are also a haunt of pseudoscientists and mystics; as the Oxford chemist Peter Atkins stated in 1976, 'the study of magnetic field effects on chemical reactions has long been a romping ground for charlatans'.[1] Indeed, all manner of exotic explanations, from telepathy and ancient ley lines (invisible pathways connecting various archaeological or geographical sites that are supposedly endowed with spiritual energy) to the concept of 'morphic resonance' invented by the controversial parapsychologist Rupert Sheldrake, have at some point been proposed as mechanisms used by migratory birds to guide them along their routes. Atkins's reservations in the 1970s were thus understandable, reflecting a scepticism prevalent among most scientists working at that time towards any suggestion that animals might be able to sense the earth's magnetic field. There just did not seem to be any molecular mechanism that would allow an animal to do so – at least, none within the realms of conventional biochemistry.

But in the same year that Peter Atkins voiced his scepticism, Wolfgang and Roswitha Wiltschko, a German husband-and-wife team of ornithologists based in Frankfurt, published a breakthrough paper in *Science*, one of the world's leading academic journals, which established beyond doubt that robins can indeed detect the earth's magnetic field.[2] More remarkably still, they showed that the birds' sense did not seem to work the way a normal compass does. For while compasses tell the difference between magnetic north and south poles, a robin could only distinguish between pole and equator.

To understand how such a compass might work we need to consider magnetic field lines, the invisible tracks that define the direction of a magnetic field and along which a compass needle will align itself when placed anywhere in that field – most familiar to us as the lines in the pattern mapped out by iron filings on a piece of paper placed above a bar magnet. Now imagine the whole earth as a giant bar magnet with the field lines emerging from its south pole and radiating outwards, curving round in loops to enter its north pole (see figure 1.1). The direction of these field lines near either pole is almost vertically into or out of the ground, but they become flatter and more nearly parallel to the surface of the planet the closer they are to the equator. So a compass that measures the angle of dip between the magnetic field lines and the surface of the earth, which we call an *inclination compass*, can distinguish between the direction towards a pole and the direction towards the equator; but it couldn't distinguish between north and south poles, since the field lines make the same angle with the ground at either end of the globe. The Wiltschkos' 1976 study established that the robin's magnetic sense worked as just such an inclination compass. The problem was that no one had a clue how any such biological inclination compass might work, because there was at that time simply no known, or even conceivable, mechanism that could account for how the angle of dip of the earth's magnetic field could be detected within an animal's body. The

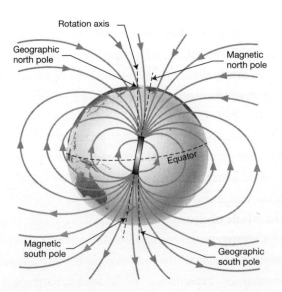

Figure 1.1: The earth's magnetic field.

answer turned out to be within one of the most startling scientific theories of modern times, and it had to do with the strange science of quantum mechanics.

A hidden spooky reality

Take a straw poll today among scientists asking them what they think is the most successful, far-reaching and important theory in the whole of science and the answer will likely depend on whether you are asking someone working in the physical or the life sciences. Most biologists regard Darwin's theory of evolution by natural selection as the most profound idea ever conceived. However, a physicist is likely to argue that quantum mechanics should have pride of place – after all, it is the foundation on which much of physics and chemistry are built and gives us a remarkably complete picture of the building blocks of the entire universe. Indeed, without its explanatory power, much of our current understanding of how the world works disappears.

Almost everyone will have heard of 'quantum mechanics', and the idea that this is a baffling and difficult area of science understood only by a tiny, very smart minority of humans is very much part of popular culture. Yet the truth is that quantum mechanics has been part of all our lives since the early twentieth century. The science was developed as a mathematical theory in the mid-1920s to account for the world of the very small (the microworld, as it's called), which is to say the behaviour of the atoms that make up everything we see around us and the properties of the even tinier particles that make up those atoms. For example, in describing the rules obeyed by electrons and how they arrange themselves within atoms, quantum mechanics underpins the whole of chemistry, material science and even electronics. Despite its strangeness, its mathematical rules lie at the very heart of most of the technological advances of the past half-century. Without quantum mechanics' explanation of how electrons move through materials, we would not have understood the behaviour of the semiconductors that are the foundation of modern electronics, and without an understanding of semiconductors we would not have developed the silicon transistor and, later, the microchip and the modern computer. The list goes on: without the advances in our knowledge thanks to quantum mechanics there would be no lasers and so no CD, DVD or blu-ray players; without quantum mechanics we would not have smartphones, satellite navigation or MRI scanners. In fact, it has been estimated that over one-third of the gross domestic product of the developed world depends on applications that would simply not exist without our understanding of the mechanics of the quantum world.

And this is just the beginning. We can look forward to a quantum future – in all likelihood within our own lifetimes – in which near-limitless electric power may become available from laser-driven nuclear fusion; when artificial molecular machines will be carrying out a vast array of tasks in the fields of engineering, biochemistry and medicine; when quantum computers will be providing artificial intelligence; and when potentially even the sci-fi technology of teleportation will be routinely used to transmit information. The twentieth century's quantum revolution is picking up pace in the twenty-first century and will transform our lives in unimaginable ways.

But what exactly *is* quantum mechanics? This is a question we will be exploring throughout this book; for a taster, we will start here with a few examples of the hidden quantum reality that underpins our lives.

Our first example illustrates one of the strange features of the quantum world, arguably its defining feature: wave–particle duality. We are familiar with the fact that we and all the things around us are composed of lots of tiny, discrete particles such as atoms, electrons, protons and neutrons. You may also be aware that energy, such as light or sound, comes as waves, rather than particles. Waves are spread out, rather than particulate; and they flow through space as – well, waves, with peaks and troughs like the waves of the sea. Quantum mechanics was born when it was discovered in the early years of the twentieth century that subatomic particles can behave like waves; and light waves can behave like particles.

Although wave–particle duality is not something you need to consider every day, it is the basis of lots of very important machines, such as the electron microscopes that allow doctors and scientists to see, identify and study tiny objects too small to show up under traditional optical microscopes, such as the viruses that cause AIDS or the common cold. The electron microscope was inspired by the discovery that electrons have wave-like properties. The German scientists Max Knoll and Ernst Ruska realized that, since the wavelength (the distance between successive peaks or troughs of any wave) associated with electrons was much shorter than the wavelength of visible light, a microscope based on electron imaging should be able to pick out much finer detail than an optical microscope. This is because any tiny object or detail that has dimensions smaller than the wave falling on it will not influence or affect the wave. Think of ocean waves with wavelengths of several metres washing up against pebbles on the beach. You would not be able to learn anything about the shape or size of an individual pebble by studying the waves. You would need much shorter wavelengths, such as those produced in a ripple tank, of the type everyone encounters in school science lessons, to 'see' a pebble by the way that waves bounce off it or diffract around it. So, in 1931, Knoll and Ruska built the world's first electron microscope and used it to take the first ever pictures of viruses, for which Ernst Ruska was awarded the Nobel Prize, perhaps rather belatedly, in 1986 (two years before he died).

Our second example is even more fundamental. Why does the sun shine? Most people are probably aware that the sun is essentially a nuclear fusion reactor that burns hydrogen gas to release the heat and sunlight that sustain all life on earth; but fewer people know that it wouldn't shine at all were it not for a remarkable quantum property that allows particles to 'walk through walls'. The sun, and indeed all stars in the universe, is able to emit these vast amounts of energy because nuclei of hydrogen atoms, each composed of just a single positively charged particle called a proton, are able to fuse, and as a result to release energy in the form of the electromagnetic radiation that we call sunlight. Two hydrogen nuclei have to be able to get very close in order to fuse; but the closer they get, the stronger the repulsive force between them becomes, as each carries a positive electric charge and 'like' charges repel. In fact, for them to get close enough to fuse, the particles have to be able to get through the subatomic equivalent of a brick wall: an apparently impenetrable energy barrier. Classical physics* – built upon Isaac Newton's laws of motion, mechanics and gravity, which describe very well the everyday world of balls, springs, steam engines (and even planets) – would predict that this shouldn't happen; particles should not be able to pass through walls and therefore the sun shouldn't shine.

But particles that obey the rules of quantum mechanics, such as atomic nuclei, have a neat trick up their sleeve: they can easily pass through such barriers via a process called 'quantum tunnelling'. And it is essentially their wave–particle duality that enables them to do this. Just as waves can flow around objects, like the pebbles on the seashore, they can also flow through objects, like the sound waves that pass through your walls when you hear your neighbour's TV. Of course, the air that carries sound waves doesn't actually pass through the walls itself: it's the vibrations in the air – sound – that cause your common wall to vibrate and push on the air in your room to transmit the same sound waves to your ear. But if you could behave like an atomic nucleus then you would sometimes be able to pass, ghost-like, straight through a solid wall.† A hydrogen nucleus in the interior of the

---

*Conventionally, the deterministic physical theories that preceded quantum mechanics, including special and general relativity, are collectively referred to as classical physics – as distinct from non-classical quantum mechanics.

†Although it would be wrong to think that quantum tunnelling entails the leaking through barriers of physical waves; rather, it is due to abstract mathematical waves that provide us with the probability of instantaneously finding the quantum particle on the other side of the barrier. We try in this book to provide intuitive analogies wherever possible to explain quantum phenomena, but the reality is that quantum mechanics is utterly counterintuitive and there is a danger of oversimplifying for the purposes of clarity.

sun manages to do precisely this: it can spread itself out and 'leak' through the energy barrier like a phantom, to get close enough to its partner on the other side of the wall to fuse. So when you are next sunning yourself on the beach, watching the waves lapping on the seashore, spare a thought for the spooky wave-like motions of quantum particles that not only allow you to enjoy the sunshine but make all life on our planet possible.

The third example is related, but illustrates a different and even weirder feature of the quantum world: a phenomenon called *superposition* whereby particles can do two – or a hundred, or a million – things at once. This property is responsible for the fact that our universe is richly complex and interesting. Not long after the Big Bang through which this universe came into being, space was awash with just one type of atom: the simplest in structure, hydrogen, which is made up of one positively charged proton and one negatively charged electron. It was a rather dull place, with no stars or planets and definitely no living organisms, because the elemental building blocks of everything around us, including us, consist of more than just hydrogen, including heavier elements such as carbon, oxygen and iron. Fortunately, these heavier elements were cooked up inside the hydrogen-filled stars; and their starting ingredient, a form of hydrogen known as deuterium, owes its existence to a bit of quantum magic.

The first step in the recipe is the one we've just described, when two hydrogen nuclei, protons, get close enough together via quantum tunnelling to release some of that energy that turns into the sunlight that warms our planet. Next, the two protons have to bind together, and this is not straightforward because the forces between them don't provide a strong enough glue. All atomic nuclei are composed of two types of particles: protons and their electrically neutral partners, neutrons. If a nucleus has too many of one type or the other, then the rules of quantum mechanics dictate that the balance has to be redressed and those excess particles will change into the other form: protons will become neutrons, or neutrons protons, via a process called beta-decay. This is precisely what happens when two protons come together: a composite of two protons cannot exist and one of them will beta-decay into a neutron. The remaining proton and the newly transformed neutron can then bind together to form an object called a deuteron (the nucleus of an atom of the heavy hydrogen isotope* called deuterium), after which further nuclear reactions enable the building of the more complex nuclei of other elements heavier than hydrogen, from helium (with two protons and either one or two neutrons) through to carbon, nitrogen, oxygen, and so on.

The key point is that the deuteron owes its existence to its ability to exist in two states simultaneously, by virtue of quantum superposition. This is because the proton and neutron can stick together in two different ways that are distinguished by how they *spin*. We will see later how this concept of 'quantum spin' is actually very different from the familiar spin of a big object, such as a tennis ball; but for now we will go with our classical intuition of a spinning particle and imagine both the proton and the neutron spinning together within the deuteron in a carefully choreographed combination of a slow, intimate waltz and a faster jive. It was discovered back in the late 1930s that within the deuteron these two particles are not dancing together in *either* one *or* the other of these two states, but in both states at the same time – they are in a blur of waltz and jive simultaneously – and it is this that enables them to bind together.†

An obvious response to this statement is: 'How do we know?' Surely, atomic nuclei are far too small to be seen, so might it not be more reasonable to assume that there is something missing in our understanding of nuclear forces? The answer is no, for it has been

---

* All chemical elements come in different varieties called isotopes. An element is defined by the number of protons in the nuclei of its atoms: hydrogen has one, helium two, and so on. But the number of neutrons the nucleus contains can vary. Thus, hydrogen comes in three varieties (isotopes): the atoms of normal hydrogen contain just a single proton, while those of the heavier isotopes, deuterium and tritium, also contain one and two neutrons, respectively.
† Technically, the deuteron owes its stability to a feature of the nuclear force that holds the proton and neutron together called the 'tensor interaction', which forces the pair to be in a quantum superposition of two angular momentum states, called S-wave and D-wave.

confirmed in many laboratories over and over again that if the proton and neutron were performing the equivalent of *either* a quantum waltz *or* a quantum jive, then the nuclear 'glue' between them would not be quite strong enough to bind them together; it is only when these two states are superimposed on top of each other – the two realities existing at the same time – that the binding force is strong enough. Think of the two superposed realities as a little like mixing two coloured paints, blue and yellow, to make a combined resultant colour, green. Although you know the green is made up of the two primary constituent colours, it is neither one nor the other. And different ratios of blue and yellow will make different shades of green. Likewise, the deuteron binds when the proton and neutron are mostly locked in a waltz, with just a tiny amount of jive thrown in.

So if particles couldn't jive and waltz simultaneously our universe would have remained a soup of hydrogen gas and nothing more – no stars would shine, none of the other elements would have formed and you would not be reading these words. We exist because of the ability of protons and neutrons to behave in this quantum counterintuitive way.

Our last example takes us back into the world of technology. The nature of the quantum world can be exploited not only to view tiny objects like viruses but also to see inside our bodies. Magnetic resonance imaging (MRI) is a medical scanning technique that generates marvellously detailed images of soft tissue. MRI scans are routinely used to diagnose disease and particularly to detect tumours inside internal organs. Most non-technical accounts of MRI avoid mentioning the fact that the technique depends on the weird way that the quantum world works. MRI uses big powerful magnets to align the axes of spinning nuclei of hydrogen atoms within the patient's body. These atoms are then zapped with a pulse of radio waves, which forces the aligned nuclei to exist in that strange quantum state of spinning in both directions at once. It is pointless even trying to visualize what this entails, because it is so far removed from our everyday experience! What is important is that when the atomic nuclei relax back to their initial state – the state they were in before they received the pulse of energy that jolted them into a quantum superposition – they release this energy, which is picked up by the electronics in the MRI scanner and used to create those beautifully detailed images of your inner organs.

So if you do ever find yourself lying in an MRI scanner, perhaps listening to music piped through your headphones, take a moment to ponder the counterintuitive quantum behaviour of subatomic particles that makes this technology possible.

Quantum biology

What does all this quantum weirdness have to do with the flight of the European robin as she navigates across the globe? Well, you will remember that the Wiltschkos' research in the early 1970s established that the robin's magnetic sense worked in the same way as an inclination compass. This was extraordinarily puzzling because, at the time, no one had a clue how a biological inclination compass might work. However, around the same time a German scientist called Klaus Schulten became interested in how electrons were transferred in chemical reactions involving free radicals. These are molecules that have lone electrons in their outer electron shell, in contrast to most electrons, which are paired up in atomic orbitals. This is important when considering that weird quantum property of spin, since paired electrons tend to spin in opposite directions, so their total spin cancels to zero. But, without a spin-cancelling twin, the lone electrons in free radicals have a net spin that gives them a magnetic property: their spin can be aligned with a magnetic field.

Schulten proposed that *pairs* of free radicals generated by a process known as a *fast triplet reaction* could have their corresponding electrons 'quantum entangled'. For subtle reasons that should become clear later on, such a delicate quantum state of the two separated electrons is highly sensitive to the direction of any external magnetic field. Schulten then went on to propose that the enigmatic avian compass might be using this kind of quantum entanglement mechanism.

We haven't mentioned quantum entanglement yet because it is probably the strangest feature of quantum mechanics. It allows particles that were once together to remain in instant, almost magical, communication with each other, despite being separated by huge distances. For example, particles that were once close but are later separated so far apart as to be located at opposite sides of the universe can, in principle at least, still be connected. In effect, prodding one particle would prompt its distant partner to jump *instantaneously*.* Entanglement was shown by the quantum pioneers to follow naturally from their equations, but its implications were so extraordinary that even Einstein, who gave us black holes and warped space-time, refused to accept it, deriding it as 'spooky action at a distance'. And it is indeed this spooky action at a distance that so often intrigues 'quantum mystics' who make extravagant claims for quantum entanglement, for example that it accounts for paranormal 'phenomena' such as telepathy. Einstein was sceptical because entanglement appeared to violate his theory of relativity, which stated that no influence or signal can ever travel through space faster than the speed of light. Distant particles should not, according to Einstein, possess instantaneous spooky connections. In this, Einstein was wrong: we now know empirically that quantum particles really can have instantaneous long-range links. But, just in case you are wondering, quantum entanglement can't be invoked to validate telepathy.

The idea that the weird quantum property of entanglement was involved in ordinary chemical reactions was considered outlandish in the early 1970s. At the time, many scientists were with Einstein in doubting whether entangled particles really existed at all, as no one had yet detected them. But over the decades since then, many ingenious laboratory experiments have confirmed the reality of these

---

* We should clarify that quantum physicists do not use this sort of simplistic language. More correctly, two distant yet entangled particles are said to be non-locally connected because they are parts of the same quantum state. But then, saying it like that doesn't help much, does it?

spooky connections; and the most famous of them was conducted as early as 1982 by a team of French physicists led by Alain Aspect at the University of Paris-South.

Aspect's team generated pairs of photons (particles of light) with entangled polarization states. Light polarization is probably most familiar to us through wearing polarized sunglasses. Every photon of light has a kind of directionality, its angle of polarization, which is a bit like the property of spin that we introduced earlier.* The photons in sunlight come with all possible polarization angles, but polarized sunglasses filter them, allowing through only those photons that have one particular polarization angle. Aspect generated pairs of photons with polarization directions that were not only different – let's say that one was pointing up and the other down – but entangled; and, like our previous dancing partners, neither of the entangled pair was actually pointing one way or another: they were both pointing in both directions simultaneously, *until they were measured.*

Measurement is one of the most mysterious – and certainly the most argued about – aspects of quantum mechanics, as it relates to the question that we are sure has occurred to you already: why don't all objects we see do all these weird and wonderful things that quantum particles can do? The answer is that, down in the microscopic quantum world, particles can behave in these strange ways, like doing two things at once, being able to pass through walls, or possessing spooky connections, only when no one is looking. Once they are observed, or measured in some way, they lose their weirdness and behave like the classical objects that we see around us. But then, of course, this only throws up another question: what is so special about measurement that allows it to convert quantum behaviour to classical behaviour?† The answer to this question is crucial to

---

* However, since light can be thought of as a wave as well as a particle, the notion of polarization (unlike quantum spin) can be more easily understood as the direction in which a light wave oscillates.

† Again, in striving for clarity we are deliberately being overly simplistic here. Measuring a certain property of a quantum particle, say its position, means we are no longer uncertain about where it is – in a sense, it is brought into focus and ceases to be fuzzy. However, this

our story, because measurement lies on the borderline between the quantum and classical worlds, the quantum edge, where we, as you will have guessed from the title of this book, are claiming life also lies.

We will be exploring quantum measurement throughout this book and we hope that you will gradually get to grips with the subtleties of this mysterious process. For now, we will just consider the simplest interpretation of the phenomenon and say that when a quantum property, such as polarization state, is measured by a scientific instrument then it is instantly forced to forget its quantum abilities, such as pointing in many directions simultaneously, and must take on a conventional classical property, such as pointing in a single direction only. So, when Aspect measured the polarization state of one of any pair of entangled photons, by observing whether it could pass through a polarized lens, it instantly lost its spooky connection with its partner and adopted just a single polarization direction. And so did its partner, instantly, no matter how far away it was; at least, that's what the equations of quantum mechanics predicted, which was of course exactly what made Einstein uneasy.

Aspect and his team carried out their famous experiment for pairs of photons that had been separated by several metres in his laboratory, far enough away that not even an influence travelling at the speed of light – and relativity tells us that nothing can travel faster than the speed of light – could have passed between them to coordinate their angles of polarization. Yet the measurements on paired particles were correlated: when one photon's polarization was pointing up, the other's was found to point down. Since 1982, the experiment has been repeated even for particles separated by

---

does not mean it now behaves like a classical particle. Due to Heisenberg's Uncertainty Principle it now no longer has a fixed velocity. Indeed, a particle in a definite position will, at that moment in time, be in a superposition of moving at all possible speeds in all possible directions. And as for quantum spin, since this property is only found in the quantum world, measuring it certainly does not make the particle behave classically.

hundreds of miles, and they still possess that spooky entangled connection that Einstein couldn't accept.

Aspect's experiment was still a few years away when Schulten proposed that entanglement was involved in the avian compass, and the phenomenon was still controversial. Also, Schulten had no idea *how* such an obscure chemical reaction could allow a robin to see the earth's magnetic field. We say 'see' here because of another peculiarity discovered by the Wiltschkos. Despite the European robin being a nocturnal migrant, activation of its magnetic compass required a small amount of light (around the blue end of the visible spectrum), hinting that the bird's eyes played a significant role in how it worked. But, aside from vision, how did its eyes also help provide a magnetic sense? With or without a radical pair mechanism, this was a complete mystery.

The theory that the avian compass had a quantum mechanism languished in the scientific back drawer for more than twenty years. Schulten moved back to the US where he set up a very successful theoretical chemical physics group at the University of Illinois at Urbana-Champaign. But he never forgot his outlandish theory, and continually rewrote a paper proposing candidate biomolecules (molecules that are made by living cells) that might generate the radical pairs necessary for the fast triplet reaction. But none really fitted the bill: either they couldn't generate radical pairs or they weren't present in birds' eyes. But in 1998 Schulten read that an enigmatic light receptor, called cryptochrome, had been found in animal eyes. This immediately set his scientific alarm bell ringing, because cryptochrome was known to be a protein that could potentially generate radical pairs.

A talented PhD student called Thorsten Ritz had recently joined Schulten's group. As an undergraduate at the University of Frankfurt, Ritz had heard Schulten give a talk on the avian compass and was hooked. When the opportunity arose, he jumped at the chance of doing a PhD in Schulten's lab, working initially on photosynthesis. When the cryptochrome story broke he shifted to working on magnetoreception, and in 2000 he wrote a paper with Schulten entitled 'A model for photoreceptor-based magnetoreception in birds', describing how cryptochrome could provide the avian eye with a quantum compass. (We will revisit this subject more fully in chapter 6.) Four years later, Ritz teamed up with the Wiltschkos to perform a study of European robins that provided the first experimental evidence in support of this theory that birds use quantum entanglement to navigate around the globe. Schulten, it seemed, had been right all along. Their 2004 paper, published in the prestigious UK-based journal *Nature*, sparked a huge amount of interest and the avian quantum compass instantly became the poster child for the new science of quantum biology.

Figure 1.2: Attendees at the 2012 Surrey workshop on quantum biology. *From left to right:* the authors, Jim Al-Khalili and Johnjoe McFadden; Vlatko Vedral, Greg Engel, Nigel Scrutton, Thorsten Ritz, Paul Davies, Jennifer Brookes and Greg Scholes.

If quantum mechanics is normal, why should we be excited about quantum biology?

We earlier described quantum tunnelling and quantum superposition both in the heart of the sun and in technological devices such as electron microscopes and MRI scanners. So why should we be surprised if quantum phenomena turn up in biology? Biology is, after all, a kind of applied chemistry, and chemistry is a kind of applied physics. So isn't everything, including us and other living creatures, just physics when you really get down to the fundamentals? This is indeed the argument of many scientists who accept that quantum mechanics must, at a deep level, be involved in biology; but they insist that its role is trivial. What they mean by this is that since the rules of quantum mechanics govern the behaviour of atoms, and biology ultimately involves the interaction of atoms, then the rules of the quantum world must also operate at the tiniest scales within biology – but *only* at those scales, with the result that they will have little or no effect on the scaled-up processes important to life.

These scientists are, of course, at least partly right. Biomolecules such as DNA or enzymes are made of fundamental particles like protons and electrons whose interactions are governed by quantum mechanics. But then, so is the structure of the book you are reading or the chair you are sitting on. The way you walk or talk or eat or sleep or even think must ultimately depend on quantum mechanical forces governing electrons, protons and other particles, just as the operation of your car or your toaster depends, ultimately, on quantum mechanics. But, by and large, you don't need to know that. Car mechanics aren't required to attend college courses on quantum mechanics, and most biology curricula don't include any mention of quantum tunnelling, entanglement or superposition. Most of us can get by without knowing that, at a fundamental level, the world operates according to an entirely different set of rules from those that we are familiar with. The weird quantum stuff that happens at the level of the very small doesn't usually make a difference to the big stuff like cars or toasters that we see and use every day.

Why not? Footballs don't pass through walls; people don't have spooky connections (despite the bogus claims of telepathy); and, sadly, you cannot be both at the office and at home at the same time. Yet the fundamental particles inside a football, or a person, can do all of these things. Why is there a fault line, an edge, between the world that we see and the world that physicists know really exists beneath its surface? This is one of the deepest problems in the whole of physics, and one that relates to the phenomenon of quantum measurement we introduced a little earlier. When a quantum system interacts with a classical measuring device, such as the polarizing lens in Alain Aspect's experiment, it loses its quantum weirdness and behaves like a classical object. But the measurements carried out by physicists cannot be responsible for the way the world we see around us appears. So what is it that carries out the equivalent quantum-behaviour-destroying function outside the physics laboratory?

The answer has to do with the way particles are arranged and how they move within large (macroscopic) objects. Atoms and molecules tend to be randomly scattered and vibrating erratically inside inanimate solid objects; in liquids and gases they are also in a constant state of random motion due to heat. These randomizing factors – scattering, vibrations and motion – cause the wavy quantum properties of particles to dissipate very quickly. So it is the combined action of all the quantum constituents of a body that performs the 'quantum measurement' on each and all of them, thereby making the world we see around us look normal. To observe the quantum weirdness you either have to go to unusual places (such as the interior of the sun), peer deep into the microworld (with instruments like electron microscopes) or carefully line up the quantum particles so that they are marching in step (as happens to the spins of hydrogen nuclei within your body when it is inside an MRI scanner – until the magnet is turned off, when the spin orientation of the nuclei is randomized again, cancelling out the quantum coherence once more). The same kind of molecular randomization is responsible for the fact that we can get by without quantum mechanics most of the time: all the quantum weirdness is washed away inside the randomly orientated and constantly moving molecular interiors of the visible inanimate objects that we see around us.

Most of the time . . . but not always. As Schulten discovered, the speed of the fast triplet chemical reaction could only be accounted for when that delicate quantum property of entanglement was involved. But the fast triplet reaction is just that: fast. And it only involves a couple of molecules. For it to be responsible for bird navigation it would have to have a lasting effect on an entire robin. So the claim that the avian magnetic compass was quantum entangled was a wholly different level of proposition from the claim that entanglement was involved in an exotic chemical reaction involving just a couple of particles; and it was met with considerable scepticism. Living cells were thought to be composed mostly of water and biomolecules in a constant state of molecular agitation that would be expected to instantly measure and scatter those weird quantum effects. By 'measure' here we do not of course mean that water molecules or biomolecules perform a measurement in the sense that we might measure the weight or the temperature of an object and then make a permanent record of this value on paper or on a computer's hard drive, or even only in our brain. What we are talking about here is what happens when a water molecule bumps into one of a pair of entangled particles: its subsequent motion will be affected by the state of that particle, so that if you were to study the water molecule's subsequent motion you could deduce some of the properties of the particle it had bumped into. So, in this sense, the water molecule has carried out a 'measurement' because its motion provides a record of the state of the entangled pair, whether or not anyone is there to examine it. This kind of *accidental* measurement is usually sufficient to destroy entangled states. So the claim that delicately arranged quantum entangled states could survive in the warm and complex interior of living cells was thought by many to be an outlandish idea, verging on madness.

Yet in recent years our knowledge of such things has made huge strides – and not only in connection with birds. Quantum phenomena such as superposition and tunnelling have been detected in lots of biological phenomena, from the way plants capture sunlight to the way that all our cells make biomolecules. Even our sense of smell or the genes that we inherit from our parents may depend on the weird quantum world. Research papers on quantum biology are now appearing regularly in the pages of the world's most prestigious scientific journals; and there exists a small but growing number of scientists who insist that aspects of quantum mechanics do indeed play a non-trivial, indeed crucial, role in the phenomenon of life, and that life is in a unique position to sustain these weird quantum properties at the edge between the quantum and classical worlds.

That these scientists are indeed few in number was made clear to us when we hosted an international workshop on quantum biology at the University of Surrey in September 2012 that was attended by most of those working in the field and managed to fit them all into a small lecture theatre. But the field is growing rapidly, driven by the excitement of discovering roles for quantum mechanics in everyday biological phenomena. And one of the most exciting areas of research – the one that might have huge implications for the development of new quantum technologies – is the recent unravelling of the mystery of how quantum weirdness manages to survive in hot, wet and messy living bodies.

But to fully appreciate the significance of these findings we must first ask a deceptively simple question: what is life?

2

What is life?

ONE OF the most successful scientific missions of all time began on 20 August 1977, when the Voyager 2 spacecraft lifted off into the Florida sky, followed two weeks later by its sister ship, Voyager 1. Two years later, Voyager 1 reached its first destination, Jupiter, where it photographed the gas giant's swirling clouds and famous great red spot before moving on to fly over the icy surface of one of its moons, Ganymede, and witness a volcanic eruption on another, Io. Meanwhile, Voyager 2 had been travelling on a different trajectory and, reaching Saturn in August 1981, began sending back stunningly beautiful photographs of the planet's rings, revealing them as a fine-braided necklace of millions of small rocks and moonlets. But nearly another decade passed before, on 14 February 1990, Voyager 1 snapped one of the most remarkable photographs ever taken: a picture of a tiny blue dot against a grainy grey background.

Over the past half-century, the Voyager missions and their fellow exploratory spacecraft have allowed humankind to walk on the moon, remotely explore the valleys of Mars, peer into the scorching deserts of Venus and even witness a comet hurtling into the gaseous atmosphere of Jupiter. But mostly, they have discovered rock . . . lots of rock. In fact, it could be argued that the exploration of our sister planetary bodies has largely been an investigation of rocks, from the ton or so of minerals brought back from the moon by the Apollo astronauts, or the microscopic fragments of comet recovered by the visit of NASA's Stardust mission, to the Rosetta probe's direct rendezvous with a comet in 2014 or the analysis of the surface of Mars by the Curiosity Rover – lots and lots of rock.

Rocks from space are of course fascinating objects, their structure and composition providing clues to the origin of the solar system, the formation of the planets and even the cosmic events that pre-date our sun's formation. But to most non-geologists, a Martian chondrite (a type of stony, non-metallic meteorite) is not so very different from a lunar troctolite (an iron- and magnesium-rich meteorite). There is, however, a place in our solar system where the basic ingredients that make up rocks and stones have been brought together in such a variety of form, function and chemistry that just one gram of the resulting material exceeds in diversity all the matter found elsewhere in the known universe. This place is, of course, that pale blue dot photographed by Voyager 1, the planet we call the earth. Most remarkably, those diverse raw materials that make our planet's surface so unique have come together to create life.

Life is remarkable. We have already discovered the amazing sense of magnetoreception possessed by the European robin, but that special skill is just one of its many and varied capabilities. It can see, smell, hear, catch flies; it can hop over the ground or between branches in a tree; and it can soar into the air to fly for hundreds of miles. Most remarkably of all, it can, with a little help from its mate, make a whole brood of similar creatures out of the same materials that make up all those rocks. And our robin is just one of the trillions of living organisms that are capable of performing scores of these and many other equally bewildering feats.

Another remarkable organism is, of course, you. Gaze up at the night sky and photons of light enter your eyes to be transmuted by retinal tissue into tiny electric currents that travel along your optic nerves to reach the nervous tissue of your brain. There they generate a flickering pattern of nerve firing that you experience as the twinkling star in the sky above you. At the same time, tiny pressure variations of less than one-billionth of atmospheric pressure are registered by the hair cell tissue of your inner ear, generating auditory nerve signals that inform you that the wind is whistling in the trees. A handful of molecules floating into your nose are picked up by specialized olfactory receptors and their chemical identity is relayed to your brain, telling you that it is summertime and that the honeysuckle is blossoming. And each tiny movement of your body, as you watch the stars, listen to the wind and sniff the air, is generated by the coordinated action of hundreds of muscles.

Yet the physical feats performed by the tissue of our own bodies, however extraordinary, pale by comparison with those executed by many of our fellow living creatures. The leafcutter ant is able to carry a load weighing thirty times its own weight, equivalent to you carrying a car on your back. And the trap-jaw ant is able to accelerate its jaws from 0 to 230 km per hour in just 0.13 milliseconds, while a Formula 1 racing car takes about forty thousand times as long (around five seconds) to reach the same speed. The Amazon electric eel can generate 600 volts of potentially lethal electricity. Birds can fly, fish can swim, worms can burrow and monkeys can swing through trees. And, as we have already discovered, many animals, including our European robin, can find their way across thousands of miles using the earth's magnetic field. For biosynthetic capability, meanwhile, nothing rivals the green variety of life on earth that bolts together molecules of air and water (plus a few minerals) to make grass, oak trees, seaweed, dandelions, giant redwoods and lichens.

All living organisms have their particular skills and specialities, such as the robin's magnetoreception or the trap-jaw ant's speedy snapping, but there is one human organ whose performance is unparalleled. The computation skill of the grey fleshy material locked within our bony skulls exceeds that of every computer on the planet and has created the Pyramids, the General Theory of Relativity, *Swan Lake*, the Rig Veda, *Hamlet*, Ming pottery and Donald Duck. And, perhaps most remarkably of all, the human brain possesses the capacity to *know* that it exists.

Yet all this diversity of living matter, with its multitudinous forms and endless variety of functions, is made up from pretty much the same atoms as those found in lumps of Martian chondrites.

The biggest question in science, one that is central to this book, is how the inert atoms and molecules found in rocks are transformed every day into running, jumping, flying, navigating, swimming, growing, loving, hating, lusting, fearing, thinking, laughing, crying, *living* stuff. Familiarity renders this extraordinary transformation unremarkable, but it is worth remembering that even in this age of genetic engineering and synthetic biology, nothing living has ever been made by humans entirely from non-living materials. That our technology has so far failed to manage a transformation that is effortlessly executed by even the simplest microbe on our planet suggests that our knowledge of what it takes to make life is incomplete. Have we overlooked some vital spark that animates the living and is absent from the non-living?

This is not to say we will be claiming that any kind of vital force, spirit or magic ingredient animates life. Our story is much more interesting than that. We will instead explore recent research showing that at least one of the missing pieces in the puzzle of life is found within the world of quantum mechanics, where objects can be in two places at once, possess spooky connections and travel through apparently impenetrable barriers. Life appears to have one foot in the classical world of everyday objects and the other planted in the strange and peculiar depths of the quantum world. Life, we will argue, lives on the quantum edge.

But can animals, plants and microbes really be governed by laws of nature that we have thus far believed to describe only the behaviour of fundamental particles? Surely living organisms made up of trillions of particles are macroscopic objects that, like footballs or cars or steam trains, should be adequately described by classical rules, such as Newton's mechanical laws or the science of thermodynamics. To discover why we need the hidden world of quantum mechanics to account for the amazing properties of living matter, we need first to embark on a short tour of science's efforts to understand what is so special about life.

The 'life force'

The central puzzle of life is this: why does matter behave so differently when it makes up a living creature compared to when it is a rock? The ancient Greeks were among the first people to attempt to probe this question. The philosopher Aristotle, probably the world's first great scientist, correctly identified certain properties of inanimate matter that were reliable and predictable: for example, the tendency of solid objects to fall, whereas fire and vapours tended to rise, and celestial objects tended to move in circular paths around the earth. But life was different: although many animals fell, they also ran; plants grew upwards and birds even flew around the earth. What made them so different from the rest of the world? An answer suggested by an earlier Greek thinker, Socrates, was recorded by his pupil Plato: 'What is it that, when present in a body, makes it living? – A soul.' Aristotle agreed with Socrates that living beings possessed souls, but he claimed that they came in different grades. The lowliest were those that inhabited plants, enabling them to grow and obtain nourishment; animal souls, one rung higher, endowed their hosts with feeling and movement; but only the human soul conferred reason and intellect. The ancient Chinese similarly believed that living beings were animated by an incorporeal life force called Qi (pronounced 'chi') that flowed through them. The concept of a soul was later incorporated into all of the major world religions; but its nature and its connection with the body remained mysterious.

Another puzzle was mortality. Souls were generally believed to be immortal, but then why is life ephemeral? The answer that most cultures came up with was that death was accompanied by departure of the animating soul from the body. As late as 1907, the American physician Duncan MacDougall claimed to be able to measure the soul by weighing his dying patients immediately before and following death. His experiments convinced him that the soul weighed about 21 grams. But why the soul had to depart from the body after the allotted three score years and ten remained an enigma.

The concept of a soul, while no longer part of modern science, did at least separate the study of the non-living from that of the living, allowing scientists to investigate the causes of motion in inanimate objects unencumbered by the questions of philosophy and theology that bedevilled any study of living creatures. The history of the study of the concept of motion is long, complicated and fascinating, but in this chapter we take you on just the briefest of tours. We have already mentioned Aristotle's view of objects possessing tendencies to move towards the earth, away from the earth or around the earth, all of which he considered to be *natural* motions. He also recognized that solid objects could be pushed, pulled and thrown, all motions that he called 'violent' and considered to be initiated by some kind of force provided by another object, such as the throwing person. But what produced the throwing motion – or the flight of a bird? There appeared to be no external cause. Aristotle claimed that living creatures, unlike inanimate objects, were capable of initiating their own motion, and that in this case the cause of such motion was the creature's soul.

Aristotle's views on the sources of motion remained predominant until the middle ages; but then something remarkable happened. Scientists (who would have described themselves at the time as natural philosophers) began to express theories about the motion of inanimate objects in the language of logic and mathematics. One could argue over who was responsible for this extraordinarily productive shift in human thought; medieval Arab and Persian scholars, such as Alhazen and Avicenna, certainly played a role, and the trend was then taken up in the emerging scholarly institutions of Europe such as the universities of Paris and Oxford. But this way of describing the world probably bore its first great fruit in the University of Padua in Italy, where Galileo enshrined simple laws of motion in mathematical formulae. In the year he died, 1642, Isaac Newton was born in Lincolnshire, England, and went on to provide an extraordinarily successful mathematical description of how the motion of inanimate objects could be changed by forces, a system which is to this day referred to as Newtonian mechanics.

Newton's forces were initially rather mysterious notions, but over the following centuries they became increasingly identified with the concept of *energy*. Moving objects were said to possess energy that could be transferred to stationary objects they bumped into, causing them to move. But forces could also be transmitted *remotely* between objects: examples of these were the gravitational force of the earth, which pulled Newton's apple to the ground, or the magnetic forces that deflected compass needles.

The incredible scientific advances initiated by Galileo and Newton gained pace in the eighteenth century, and by the close of the nineteenth the basic framework of what came to be known as *classical physics* was pretty much established. By this time, it was known that other forms of energy, such as heat and light, were also capable of interacting with the constituents of matter, atoms and molecules, causing them to become hotter, emit light or change colour. Objects were considered to be composed of particles whose motion was controlled by the forces of gravity or electromagnetism.* So the material world, or at least the inanimate objects in it, was divided into two distinct entities: visible matter, composed of particles, and the invisible forces that acted between them in an as yet poorly understood way, either as waves of energy propagating through space or in terms of force fields. But what about the animate matter that made up living organisms? What was it made of and how did it move?

---

*In the late nineteenth century, the Scottish physicist James Clerk Maxwell demonstrated that the electrical and magnetic forces were two facets of the same electromagnetic force.

Triumph of the machines

The ancient idea that all living creatures were animated by some kind of supernatural substance or entity did at least provide some kind of explanation for the remarkable differences between the living and the non-living. Life was different because it was moved by a spiritual soul rather than by any of those mundane mechanical forces. But this was always an unsatisfactory explanation – equivalent to accounting for the motion of the sun, moon and stars by claiming that they were pushed around by angels. In truth, there was no real explanation, as the nature of souls (and angels) remained entirely mysterious.

In the seventeenth century, the French philosopher René Descartes provided a radical alternative view. He was impressed by the mechanical clocks, toys and automata dolls that provided amusement for the courts of Europe at the time, and was inspired by their mechanisms to make the revolutionary claim that the bodies of plants and animals, including humans, were merely elaborate machines composed of conventional materials and driven by mechanical devices such as pumps, cogs, pistons and cams that were in turn subject to those same forces that governed the motion of inanimate matter. Descartes exempted the human mind from his mechanistic view, leaving it with an immortal soul; but his philosophy did at least attempt to provide a scientific framework that accounted for life in terms of the physical laws that were being discovered to govern inanimate objects.

The mechanistic biological approach was continued by a near-contemporary of Sir Isaac Newton's: the physician William Harvey, who discovered that the heart was nothing more than a mechanical pump. A century later, the French chemist Antoine Lavoisier demonstrated that a respiring guinea pig consumes oxygen and generates carbon dioxide, just like the fire that provided the motive force of the new technology of steam engines. He accordingly concluded that 'respiration is thus a very slow combustion phenomenon, very similar to that of coal'. As Descartes might have predicted, animals appeared not to be so very different from the coal-powered locomotives that were soon hauling the industrial revolution across Europe.

But can the forces that move steam trains also move life? To answer that question, we need to understand how steam trains climb up hills.

A molecular billiard table

The science of how heat interacts with matter is called *thermodynamics*; and its key insight was provided by the nineteenth-century Austrian physicist Ludwig Boltzmann, who took the bold step of treating the particles of matter rather like a very large collection of randomly colliding billiard balls that obeyed Newton's mechanical laws.

Imagine the surface of a billiard table* divided into two sides by a movable baton. All the balls, including the cue ball, are on the left-hand side of the baton, with the pack neatly arranged in a triangle. Now imagine hitting the pack very hard with the cue ball so that the balls are set in rapid motion in all directions, colliding into each other and bouncing off the rigid walls of the table and the movable baton. Consider what happens to the baton: it will be subjected to the force of many collisions coming from the left, where all the balls are, but no collisions coming from the empty side of the table on the right. Despite the motion of the balls being entirely random, the baton, driven by all those randomly moving balls, will experience an average force that pushes it to the right, expanding the playing area on the left and contracting the empty area. We could further imagine harnessing our billiard table to do some work by constructing a contraption of levers and pulleys that would capture the baton's rightward motion and redirect it to, say, push a toy train up a toy hillside.

This, Boltzmann realized, is essentially how heat engines push real steam locomotives – remember this was the age of steam – up real

---

* We are talking here about the American game of billiards or pool.

hillsides. The water molecules inside the cylinder of the steam engine behave much like the billiard balls after being scattered by the impact of the cue ball: their random motion is speeded up by the heat of the furnace so that the molecules bump into each other, and the engine piston, more energetically, pushing the piston outwards to drive the shafts, gears, chains and wheels of the steam train and thereby deliver a directed motion. A century and more after Boltzmann, your own petrol-powered automobile works by precisely the same principles, but with the products of petrol combustion replacing steam.

A remarkable aspect of the science of thermodynamics is that this really is all there is to it. The orderly motion of every heat engine that has ever been built is delivered by harnessing the average motion of trillions of randomly moving atoms and molecules. Not only that, but the science is extraordinarily general, applicable not only to heat engines, but to nearly all the standard chemistry that takes place whenever we burn coal in air, allow an iron nail to rust, cook a meal, manufacture steel, dissolve salt in water, boil a kettle or send a rocket to the moon. All these chemical processes involve the exchange of heat and they are, at a molecular level, all driven by thermodynamic principles that are based on random motion. In fact, almost all of the non-biological (physical and chemical) processes that cause change in our world are driven by thermodynamic principles. Ocean currents, violent storms, the weathering of rocks, the burning of forests and the corrosion of metals are all controlled by the inexorable forces of chaos that underpin thermodynamics. Each complex process may appear structured and orderly to us, but at their core they are all driven by random molecular motion.

Life as chaos?

Is the same then true of life? Let us return to our billiard table, but at the beginning of the game with the balls now arranged in a neat triangle. This time, we also add a large number of extra balls (let's imagine this is a very big table) and arrange for them to be knocked violently around the triangle of original balls. Once again, the random collision-driven motion of the dividing baton will be harnessed to do useful work; but instead of allowing it simply to drive a toy train up a hillside, we will engineer an even cleverer device. This time our motion-driven machine, impelled by the chaotic bouncing around of all those balls, will do something rather special: it will maintain the neat triangle of original balls in among the chaos. Every time one of the balls from the triangular assembly is knocked out of its position by a randomly moving ball, some kind of sensory device detects the event and directs a mechanical arm to replace the missing ball in the triangle – maybe filling a gap in one of its corners – with an identical one drawn from all the randomly colliding balls.

We hope you can see that the system is now using some of the energy made available by all those random molecular collisions to maintain part of itself in a highly ordered state. In thermodynamics, the term *entropy* is used to describe a lack of order, and so highly ordered states are described as having low entropy. Our billiard table can be said to be harvesting energy from high-entropy (chaotic) collisions to maintain part of itself, the triangle of balls in the middle, in a low-entropy (ordered) state.

Never mind for now how we might engineer such a tricky contrivance: the key point is that our entropy-driven billiard table is doing something very interesting. With only chaotic ball motion to work with, this new system of balls, table, baton, ball detection device and movable arm is able to keep order in a subsystem of itself.

Now let's imagine another level of sophistication: this time, some of the available energy from the moving baton – we could call this the system's *free energy** – is used to *construct* and *maintain* the sensory device and movable arm and even to use lots of the billiard balls as raw material to build these devices in the first place. Now the entire system is self-sustaining and could in principle, so long as it is

---

* 'Free energy' is one of the most important concepts in thermodynamics and corresponds pretty well to the description presented here.

continually supplied with lots of randomly moving balls and enough space for the baton to move, maintain itself indefinitely.

Finally, as well as maintaining itself, this extended system will accomplish one additional and amazing feat: it will use the available free energy to detect, capture and arrange billiard balls to make a copy of itself in its entirety: the table, the baton, the ball-detection device and the movable arm as well as the triangle of balls. And these copies would similarly be able to harness *their* billiard balls and the free energy available from their collisions to make more of such self-sustaining devices. And these copies . . .

Well, you'll have guessed where this is going. Our imaginary DIY project has constructed a billiard-ball-driven equivalent of life. Just like a bird, a fish or a human, our imaginary device is able to sustain and replicate itself by harvesting free energy from random molecular collisions. And although this is a complex and difficult task, its driving force is generally considered to be exactly the same as that used for pushing steam trains up hillsides. In life, billiard balls are replaced by molecules obtained from food, but although the process is far more complex than that described in our simple example, the principle is the same: free energy harvested from random molecular collisions (and their chemical reactions) is directed to maintain a body and make a copy of that body.

Is life, then, just a branch of thermodynamics? When we are out on a hike, do we ascend hillsides through the same processes that push steam locomotives? And is the robin's flight not so very different from that of a cannonball? When it comes down to it, is life's vital spark just random molecular motion? To answer this question, we need to take a closer look at the fine structure of the living.

Peering deeper into life

The first major advance in uncovering the fine structure of life was provided by the seventeenth-century 'natural philosopher' Robert Hooke, who peered down his rudimentary microscope and saw

what he called 'cells' in thin slices of cork, and by the Dutch micros-
copist Anton van Leeuwenhoek, who identified what he called
'animalcules' – now referred to as unicellular life – in drops of pond
water. He also observed plant cells, red blood cells and even sperma-
tozoa. It was later understood that all living tissue was divided into
these cellular units, the building blocks of living bodies. The German
physician and biologist Rudolf Virchow wrote in 1858:

> Just as a tree constitutes a mass arranged in a definite manner, in
> which in every single part, in the leaves as in the root, in the trunk
> as in the blossom, cells are discovered to be the ultimate elements,
> so is it also with the forms of animal life. Every animal presents
> itself as a sum of vital entities, every one of which manifests all the
> characteristics of life.

As living cells were studied in ever greater detail by more powerful microscopes, their internal structure was revealed to be highly
complex, each with a nucleus in the centre filled with chromosomes
and surrounded by *cytoplasm* in which were embedded specialized
subunits called *organelles* that, like our body's organs, perform particular functions within the cell. For example, an organelle called the
mitochondrion performs respiration inside human cells, whereas
the chloroplast organelle performs photosynthesis inside plant cells.
Overall, the cell gives the impression of a busy miniature manufacturing plant. But what keeps it going? What *animates* the cell? Initially,
cells were generally thought to be filled with 'vital' forces, essentially
equivalent to Aristotle's concept of the soul; and for much of the
nineteenth century, the belief in vitalism – that living creatures were
animated by a force absent from the non-living – persisted. Cells
were thought to be filled with a mysterious living substance called
protoplasm that was described in almost mystical terms.

But vitalism was undermined by the work of several nineteenth-
century scientists who succeeded in isolating chemicals from living

cells that were identical to those synthesized in the laboratory. For example, in 1828 the German chemist Friedrich Wöhler managed to synthesize urea, a biochemical that had previously been thought to be peculiar to living cells. Louis Pasteur even succeeded in reproducing chemical transformations, such as fermentation, previously thought to be unique to life, by using extracts from living cells (later called enzymes). Increasingly, the matter of the living appeared to be made up from pretty much the same chemicals that made up the non-living, and thereby likely to be governed by the same chemistry. Vitalism gradually gave way to mechanism.

By the end of the nineteenth century, the biochemists had pretty much triumphed over the vitalists.* Cells were considered to be bags of biochemicals operated by a complex chemistry, but one that was nevertheless based on the random billiard-ball-like molecular motion described by Boltzmann. Life, it was generally believed, was indeed just elaborate thermodynamics.

Except for one aspect – arguably the most important of all.

Genes

The ability of living organisms to faithfully transmit the instructions to make another of themselves – whether a robin, rhododendron or a person – was, for centuries, profoundly puzzling. In his '51st Exercitation' of 1653, the English surgeon William Harvey wrote:

> Although it be a known thing subscribed by all, that the foetus assumes its origin and birth from the male and female, and consequently that the egge is produced by the cock and henne and the chicken out of the egge, yet neither the schools of physicians nor Aristotle's discerning brain have disclosed the manner how the cock and its seed doth mint and coine the chicken out of the egge.

* Although it should be made clear that some biochemists were also vitalists.

Part of the answer was provided two centuries later by the Austrian monk and plant scientist Gregor Mendel, who around 1850 was breeding peas in the garden of the Augustinian abbey at Brno. His observations led him to propose that traits such as flower colour or pea shape were controlled by heritable 'factors' that could be transmitted, unchanged, from one generation to the next. Mendel's 'factors' thereby provided a repository of heritable information that allowed peas to retain their character through hundreds of generations – or through which 'the cock and its seed doth mint and coine the chicken out of the egge'.

Mendel's work was famously overlooked by most of his contemporaries, including Darwin, and it wasn't until the early twentieth century that it was rediscovered. His factors were renamed *genes* and were soon incorporated into the growing mechanistic consensus of twentieth-century biology. But although Mendel had shown that these entities must exist inside living cells, nobody had ever seen them or knew what they were composed of. However, in 1902 the American geneticist Walter Sutton noted that intracellular structures called *chromosomes* tended to follow the inheritance of Mendelian factors, leading him to propose that genes were located in chromosomes.

But chromosomes are big (relatively speaking) and complicated structures composed of protein, sugars and a biochemical called deoxyribonucleic acid, or DNA. It wasn't initially clear which, if any, of these components was responsible for heredity. Then, in 1943, the Canadian scientist Oswald Avery managed to transfer a gene from one bacterial cell to another by extracting DNA from the donor cell and injecting it into the recipient cell. The experiment demonstrated that it was the DNA in the chromosomes that carried all the vital genetic information, not the proteins or other biochemicals.* Nevertheless, there seemed to be nothing magical about DNA; at this point, it was considered just an ordinary chemical.

---

* At the time, however, Avery's experiments were not considered definitive proof that DNA was the genetic material – that debate was still rumbling away in Crick and Watson's time.

And yet the question remained: how did this all work? How does a chemical deliver the information needed to provide 'the manner how the cock and its seed doth mint and coine the chicken out of the egge'? And how were genes copied and replicated from one generation to the next? Conventional chemistry, driven by those Boltzmann ball-like molecules, just didn't seem capable of providing the means to store, copy and accurately transmit genetic information.

The answer was famously provided in 1953 when James Watson and Francis Crick, working in the Cavendish Laboratory in Cambridge, managed to fit a remarkable structure to the experimental data obtained from DNA by their colleague Rosalind Franklin: the double helix. Each DNA strand was found to be a kind of molecular string made up of atoms of phosphorus, oxygen and a sugar called deoxyribose, with chemical structures called *nucleotides** strung out like beads on that string. These nucleotide beads come in four varieties: adenine (A), guanine (G), cytosine (C) and thymine (T), so their arrangement along the DNA strand provides a one-dimensional sequence of genetic letters such as 'GTCCATTGCCCGTATTACCG'. Francis Crick had spent the war years working at the Admiralty (the authority responsible for the command of the Royal Navy), so it's conceivable he may have been familiar with codes, such as those produced by the German Enigma machines that were being decoded at Bletchley Park. In any case, when he saw the DNA strand he immediately recognized it as a code, a sequence of information that provided the crucial instructions of heredity. And, as we will discover in chapter 7, identification of the double helical DNA strand also solved the problem of how genetic information is copied. At a stroke, two of the greatest mysteries of science had been solved.

The discovery of the structure of DNA provided a mechanistic key that unlocked the mystery of genes. Genes are chemicals and

---

* These chemical structures consist of nucleotide bases made up of carbon, nitrogen, oxygen, hydrogen, along with at least one phosphate group, which are chemically bolted onto the DNA strand.

chemistry is just thermodynamics; so did the discovery of the double helix finally bring life entirely into the realm of classical science?

Life's curious grin

In Lewis Carroll's *Alice's Adventures in Wonderland*, the Cheshire cat has a habit of disappearing, leaving only his grin, prompting Alice to remark that she has 'often seen a cat without a grin but never a grin without a cat'. Many biologists experience similar bemusement when, despite knowing how thermodynamics operates in living cells and how genes encode everything that is required to form the cell, the mystery of what life really is continues to grin back at them.

One problem is the sheer complexity of biochemical reactions going on inside every living cell. When chemists artificially produce an amino acid or a sugar they almost always synthesize only a single product at a time, which they manage by carefully controlling the experimental conditions for the selected reaction, such as temperature and the concentrations of the various ingredients, to optimize the synthesis of their target compound. This is not an easy task and requires careful control of many different conditions inside customized flasks, condensers, separation columns, filtration devices and other elaborate chemical apparatus. Yet every living cell in your body is continually synthesizing thousands of distinct biochemicals within a reaction chamber filled with just a few millionths of a microlitre of fluid.* How do all those diverse reactions proceed concurrently? And how is all this molecular action orchestrated within a microscopic cell? These questions are the focus of the new science of *systems biology*; but it is fair to say that the answers remain mysterious!

Another puzzle of life is mortality. A characteristic of chemical reactions is that they are always reversible. We may write a chemical reaction in the direction: substrates $\rightarrow$ products. But, in reality, the

---

* One microlitre of water has a volume of one cubic millimetre.

reverse reaction: product $\rightarrow$ substrate, is also always proceeding simultaneously. It's just that, under a given set of conditions, one direction tends to dominate. However, it is always possible to find another set of conditions that favours the reverse chemical direction. For example, when fossil fuels burn in air, the substrates are carbon and oxygen and the sole product is the greenhouse gas carbon dioxide. This is normally considered to be an irreversible reaction; but some forms of carbon capture technology are working towards reversing that process by using a source of energy to drive the reaction backwards. For example, Rich Masel from Illinois University has set up a company, Dioxide Materials, which aims to use electricity to convert atmospheric carbon dioxide into vehicle fuel.[1]

Life is different. No one has ever discovered a condition that favours the direction: dead cell $\rightarrow$ live cell. This was of course the puzzle that prompted our ancestors to come up with the idea of a soul. We no longer believe that a cell possesses any kind of soul; but what is it then that is irrevocably lost when a cell or a person dies?

At this point you might be thinking: what about that newly heralded science of synthetic biology? Surely the practitioners of that science must possess the key to life's mystery? Probably synthetic biology's most famous practitioner is the genome-sequencing pioneer Craig Venter, who in 2010 conjured up a scientific storm when he claimed to have created *artificial life*. His work made headlines across the world and sparked fears of new races of artificially grown creatures taking over the planet. But Venter and his team managed only to modify an existing life form, rather than truly creating new life. They did this by first synthesizing DNA encoding the entire genome of a bacterial pathogen, known as *Mycoplasma mycoides*, that causes a disease in goats. They then injected their synthesized DNA genome into a living bacterial cell and very cleverly managed to persuade it to replace its original (and only) chromosome with their synthesized version.

This work was undoubtedly a technical *tour de force*. The bacterial chromosome contains 1.8 million genetic letters that all had to be strung together in precisely the right sequence. But, in essence, what the scientists had done was to perform the same transformation that all of us effortlessly manage when we convert the inert chemicals in our food into our own living flesh.

Venter and his team's successful synthesis and insertion of a substitute bacterial chromosome opens up an entirely new field of synthetic biology that we will be revisiting in the final chapter. It is likely to yield more efficient ways to make drugs, grow crops or destroy pollutants. But in these and many other similar experiments, scientists did not create new life. Despite Venter's achievement, life's essential mystery continues to grin back at us. The Nobel Prize winning physicist Richard Feynman is credited with insisting that 'what we can't make, we don't understand'. By this definition, we do not understand life because we have not yet managed to make it. We can mix biochemicals, we can heat them, we can irradiate them; we can even, like Mary Shelley's Frankenstein, use electricity to animate them; but the only way we can make life is by injecting these biochemicals into already living cells, or by eating them, thereby making them part of our own bodies.

So why is it that we are still unable to perform a trick that is effortlessly executed by trillions of the lowliest microbes every second? Are we missing an ingredient? This is the question that a famous physicist, Erwin Schrödinger, pondered more than seventy years ago; and his very surprising answer is central to the theme of this book. To understand why Schrödinger's solution to life's deepest mysteries was and continues to be so revolutionary we need to return to the beginning of the twentieth century, before the double helix had been discovered, when the world of physics was being turned upside down.

The quantum revolution

The explosion of scientific knowledge during the Enlightenment of the eighteenth and nineteenth centuries produced Newtonian mechanics, electromagnetism and thermodynamics, and showed that together these three areas of physics successfully described the motion and behaviour of all macroscopic everyday objects and phenomena in our world, from cannonballs to clocks, from storms to steam trains, from pendulums to planets. But in the late nineteenth and early twentieth centuries, when physicists turned their attention to the microscopic constituents of matter – atoms and molecules – they discovered that the familiar laws no longer applied. Physics needed a revolution.

The first major breakthrough – the concept of the 'quantum' – was made by the German physicist Max Planck, who presented his results in a seminar to the German Physical Society on 14 December 1900, a date widely regarded as the birthday of quantum theory. The conventional understanding at the time was that heat radiation travelled, like other forms of energy, through space as a wave. The problem was that the wave theory could not explain the way certain hot objects radiate energy. So Planck proposed the radical idea that the matter in the walls of these hot bodies vibrated at certain discrete frequencies, which had the consequence that the heat energy was only radiated in tiny discrete lumps, or 'quanta', that could not be subdivided. His simple theory was remarkably successful, but was a radical departure from the *classical* theory of radiation, in which energy was regarded as continuous. His theory suggested that energy, instead of flowing out of matter like water pouring continuously from a tap, came out as a collection of separate, indivisible packages – as if from a slowly dripping tap.

Planck was never comfortable with the idea that energy was lumpy, but five years after he proposed his quantum theory, Albert Einstein extended this idea and suggested that all electromagnetic radiation, including light, is 'quantized' rather than continuous, coming in discrete packets, or particles, which we now call photons. He proposed that this way of thinking about light could account for a long-standing puzzle known as the photoelectric effect, a phenomenon whereby light could knock electrons out of matter. It was this work, rather than his more famous theories of relativity, that would win Einstein the Nobel Prize in 1921.

But there was also plenty of evidence that light behaves as a spread-out and continuous wave. So how can light be both lumpy and wavy? It didn't seem to make sense at the time; at least, not within the framework of classical science.

The next giant step was taken by the Danish physicist Niels Bohr, who turned up in Manchester in 1912 to work with Ernest Rutherford. Rutherford had just proposed his famous planetary model of the atom, consisting of a tiny dense nucleus at the centre, surrounded by even tinier orbiting electrons. But nobody understood how atoms remained stable. According to standard electromagnetic theory, the negatively charged electrons would constantly emit light energy as they orbited the positively charged nucleus. In doing so, they would lose energy and very quickly (within a thousand billionth of a second) spiral inwards towards the nucleus, causing the atom to collapse. But electrons don't do this. So what was their trick?

To explain the stability of atoms, Bohr proposed that electrons aren't free to occupy any orbit around the nucleus, but instead only certain fixed ('quantized') orbits. An electron can only drop to the next lower orbit by emitting a lump, or quantum, of electromagnetic energy (a photon) of exactly the same value as the difference in energies between the two orbits involved. Likewise, it can only jump to a higher orbit by absorbing a photon of the appropriate energy.

A way to visualize this difference between classical and quantum theory, and explain why the electron would occupy only certain fixed orbits in the atom, is to compare how notes are played on a guitar and a violin. When a violinist plays a note she or he presses a finger onto one of the strings anywhere along the neck of the violin to shorten that string and thereby obtain the note when the bow drags across it, making it vibrate. Shorter strings vibrate at high frequencies (lots of vibrations per second) to generate higher notes whereas longer strings vibrate at low frequencies (few vibrations per second) to generate lower notes.

Before going on we need to say a few words about one of the fundamental features of quantum mechanics, which is the way frequency and energy are intimately related.* We saw in the last chapter how subatomic particles also have wave properties, which means that, like any spread-out wave, they have a wavelength and oscillation frequency associated with them. Fast vibrations or oscillations are always more energetic than slow vibrations – think of your spin dryer which must spin (oscillate) at high frequency in order to possess sufficient energy to force the water out of your clothes.

Back to our violin. The pitch of the note (its vibrational frequency) can vary continuously, depending on the length of string between its fixed end and the player's finger. This is equivalent to a classical wave that can take on any wavelength (the distance between successive peaks). We will therefore define the violin as a *classical* instrument – not in the 'classical music' sense, but rather in the non-quantized classical physics sense. Of course, this is why it is so hard to play the violin well, because the musician must know precisely where to place their finger to get just the right note.

But the neck of a guitar is different; it has 'frets' at intervals along its length – spaced metal bars raised slightly above the neck, but not touching the strings above them. So when a guitarist places his or her finger on a string, it is pushed down onto the fret making that, rather than the finger, temporarily one end of the string. When the string is plucked, the pitch of the resulting note is produced by the string's vibration only between the fret and the bridge. The finite number of frets means that only certain, discrete, notes can be played on the guitar. Adjusting the position of the finger between two frets will not alter the note when that string is plucked. The guitar is thus akin to a *quantum* instrument. And since, according to quantum theory, frequency and energy are related, the vibrating guitar string must possess discrete, rather than continuous, energies. In a similar way, fundamental particles, such as electrons, can only be associated with certain characteristic wave frequencies, each associated with its own

---

* In fact, the relationship is encapsulated in the equation proposed by Max Planck in 1900. It is written as $E = h\nu$, where E is energy, ν is frequency and h is called Planck's constant. You can see from this equation that energy is proportional to frequency.

discrete energy level. When it jumps from one energy state to another it must absorb or emit radiation corresponding to the energy difference between the level from which it jumps and the level at which it lands.

By the mid-1920s, Bohr, now back in Copenhagen, was one of several European physicists working feverishly on a more complete and coherent mathematical theory to describe what was going on in the subatomic world. One of the most brilliant of this group was a young German genius, Werner Heisenberg. While recovering from a bout of hay fever on the German island of Heligoland during the summer of 1925, Heisenberg made a major advance in formulating the new mathematics needed to describe the world of atoms. But it was a strange kind of mathematics, and what it told us about atoms was even stranger. For instance, Heisenberg argued not only that we could not say exactly where an atomic electron was if we weren't measuring it, but that the electron itself did not have a definite location because it was spread out in a fuzzy, unknowable way.

Heisenberg was forced to conclude that the atomic world is a ghostly, insubstantial place that crystallizes into sharp existence only when we set up a measuring device to interact with it. This is the quantum measurement process that we briefly described in the last chapter. Heisenberg showed that this process reveals only those features that it is specifically designed to measure – much as the individual instruments on the dashboard of a car each give information about just one aspect of its operation, such as its speed, the distance travelled or the temperature of the engine. Thus we could set up an experiment to determine the precise position of an electron at some given time; we could also set up a different experiment to measure the speed of the same electron. But Heisenberg showed mathematically that it is impossible to set up a single experiment in which we can measure, as accurately as we wish, both where an electron is and how fast it is moving, simultaneously. In 1927 this concept became encapsulated in the famous Heisenberg Uncertainty Principle, which has since been verified many thousands of times in laboratories around the world. It remains one of the most important ideas in the whole of science and one of the foundation stones of quantum mechanics.

In January 1926, at much the same time that Heisenberg was developing his ideas, the Austrian physicist Erwin Schrödinger wrote a paper outlining a very different picture of the atom. In it he proposed a mathematical equation, now known as the Schrödinger equation, which describes not the way a particle moves but the way that a wave evolves. It suggested that rather than an electron being a fuzzy particle in the atom, with an unknowable position as it orbits the nucleus, it is instead a wave spread throughout the atom. Unlike Heisenberg, who believed that it is impossible to have a picture of an electron at all when we are not measuring it, Schrödinger preferred to think of it as a real physical wave when we aren't looking at it, which 'collapses'* to a discrete particle whenever we do look. His version of atomic theory became known as *wave mechanics* and his famous equation describes how these waves evolve and behave over time. Today we regard both Heisenberg's and Schrödinger's descriptions as different ways of interpreting the mathematics of quantum mechanics and both, each in its own way, as correct.

Schrödinger's wave function

When we wish to describe the motion of everyday objects, whether cannonballs or steam trains or planets, each one composed of trillions of particles, we solve the problem using a set of mathematical equations that date back to the work of Isaac Newton. But if the system we are describing resides in the quantum world, then we have to use Schrödinger's equation instead. And here lies the profound difference between the two approaches, for in our Newtonian world the solution of an equation of motion is a number, or a set of

---

* This process is sometimes called 'collapse of the wave function', and in modern standard textbooks it refers to a change to the mathematical description of the electron rather than a physical collapse of a real wave.

numbers, that define(s) the precise location of an object at a given moment in time. In the quantum world, the solution of the Schrödinger equation is a mathematical quantity called the wave function, which does *not* tell us the precise location of, say, an electron at a particular moment in time, but instead provides a whole set of numbers that describe the *likelihood* of the electron's being found at different locations in space *if we were to look for it there.*

Of course, your first reaction to this should be: but this is not good enough; just telling us where the electron *might* be does not sound like very useful information. You will want to know exactly where the particle *is.* But unlike a classical object that always occupies a definite position in space, an electron could be in multiple places at once until the moment it is measured. The quantum wave function is spread out over all space – meaning that in describing an electron, say, the best we can do is work out a set of numbers that give the probability of finding it not at a single location, but at every point in space simultaneously. It is important to realize, however, that these quantum probabilities do not represent some deficiency in our knowledge that could be cured by obtaining more information; rather, they are a fundamental feature of the natural world at this microscopic scale.

Imagine a jewellery thief has just been given parole and is released from prison. Instead of mending his ways, he immediately reverts to his old habits and starts breaking into houses all over town. By studying a map, the police are able to trace his likely whereabouts from the moment he is freed. While they cannot pinpoint his exact location at any given time, they can assign probabilities to burglaries being committed by him in various districts.

To begin with, the houses close to the prison are most at risk, but in time the area under threat grows larger. And, knowing the kind of properties he has targeted in the past, the police are also able to say with some confidence that the wealthier districts, with their higher-value jewellery, are more at risk than the poorer ones. This one-man crime wave spreading through the city can be thought of as a wave of probability. It is not tangible and it is not real, just a set of abstract numbers that can be assigned to the various parts of the city. In a similar way, a wave function spreads out from the point where an electron was last seen. Calculating the value of this wave function at different positions and times allows us to assign probabilities to where it might show up next.

Now, what if the police act on a tip-off and are able to catch the thief red-handed as he crawls out of a window with his bag of 'swag' over his shoulder? Immediately, their spread-out probability distribution describing the thief's whereabouts has collapsed to being definitely at one location and definitely not anywhere else. Likewise, if the electron is detected in a certain location then its wave function is instantly altered. At the moment of detection there will be zero probability of finding it anywhere else.

However – and here is where the analogy breaks down – even though before they catch him, the police can only assign probabilities to the whereabouts of the burglar, they know this is only due to their lack of information. After all, the burglar has not actually spread himself across the city, and while the police must regard him as potentially being anywhere, in reality he is of course only ever in one place at any given time. But, in stark contrast to the burglar, when we are not tracking the motion of an electron we cannot assume it nevertheless exists in some definite place at some particular time. Instead, all we have to describe it is the wave function, which is everywhere at once. Only through the act of looking (carrying out a measurement) can we 'force' the electron to become a localized particle.

By 1927, thanks to the efforts of Heisenberg, Schrödinger and others, the mathematical underpinnings of quantum mechanics were essentially complete. Today, they constitute the foundation on which much of physics and chemistry are built and give us a remarkably complete picture of the building blocks of the entire universe.

Indeed, without the explanatory power of quantum mechanics in describing how everything fits together, much of our modern technological world would simply not be possible.

So it was that in the late 1920s, flushed with their recent successes in taming the atomic world, several of the quantum pioneers strode out of their physics laboratories to conquer a different area of science: biology.

The early quantum biologists

In the 1920s, life was still a mystery. Although nineteenth-century biochemists had made great advances in constructing a mechanistic understanding of the chemistry of life, many scientists continued to cling to the vitalist principle that biology could not be reduced to chemistry and physics but required its own set of laws. The 'protoplasm' inside living cells was still considered a mysterious form of matter animated by unknown forces, and the secret of heredity continued to elude the growing science of genetics.

But during that decade there emerged a new breed of scientists, known as organicists, who rejected the ideals of both the vitalists and the mechanists. These scientists accepted that there was something mysterious about life, but claimed that the mystery *could*, in principle, be explained by yet-to-be-discovered laws of physics and chemistry. One of the greatest proponents of the organicist movement was another Austrian, the exotically named Ludwig von Bertalanffy, who wrote some of the earliest papers on theories of biological development and highlighted the need for some new biological principle to describe the essence of life in his 1928 book *Kritische Theorie der Formbildung* (*Critical Theory of Morphogenesis*). His ideas, and in particular that book, influenced many scientists, among them another pioneering quantum physicist, Pascual Jordan.

Born and educated in Hanover, Pascual Jordan studied under

one of the founding fathers of quantum mechanics, Max Born,* in Göttingen, Germany. In 1925 Jordan and Born published the classic paper 'Zur Quantenmechanik' ('On quantum mechanics'). A year later, a 'sequel', 'Zur Quantenmechanik II', was published by Jordan, Born and Heisenberg. Known as the *Dreimännerwerk*, this 'three-man paper' is regarded as one of the classics of quantum mechanics, for it took Heisenberg's remarkable breakthrough and developed it into a mathematically elegant way of describing the behaviour of the atomic world.

The following year, Jordan did what any self-respecting young European physicist of his generation would have done if given the chance: he spent time in Copenhagen working with Niels Bohr. Some time around 1929, the two men began discussing whether quantum mechanics might have some application in the field of biology. Pascual Jordan returned to Germany, to a post at the University of Rostock, from where over the next couple of years he maintained a correspondence with Bohr about the relationship between physics and biology. Their ideas culminated in what is arguably the first scientific paper on quantum biology, written by Jordan in 1932 for the German journal *Die Naturwissenschaften* and entitled 'Die Quantenmechanik und die Grundprobleme der Biologie und Psychologie' ('Quantum mechanics and the fundamental problems of biology and psychology').[2]

Jordan's writings do contain several interesting insights into the phenomenon of life; however, his biological speculations became increasingly politicized and aligned with Nazi ideology, even claiming that the concept of a single dictatorial leader (*Führer*) or guide was a central principle of life.

> We know that there are in a bacterium, among the enormous number of molecules constituting this . . . creature . . . a very small number of special molecules endowed with dictatorial authority

---

* It was Max Born who first made the connection between Schrödinger's wave function and probabilities in quantum mechanics.

over the total organism; they form a *Steuerungszentrum* [steering centre] of the living cell. Absorption of a light quantum anywhere outside of this *Steuerungszentrum* can kill the cell just as little as a great nation can be annihilated by the killing of a single soldier. But absorption of a light quantum in the *Steuerungszentrum* of the cell can bring the entire organism to death and dissolution – similar to the way a successfully executed assault against a leading [*führenden*] statesman can set an entire nation into a profound process of dissolution.[3]

This attempt to import Nazi ideology into biology is both fascinating and chilling. But there is within it the germ of a curious idea, what Jordan called *Verstärkertheorie*, or amplification theory. Jordan pointed out that inanimate objects were governed by the average random motion of millions of particles, such that the motion of a single molecule has no influence whatsoever on the whole object. But life, he argued, was different, because it was ruled by a very few molecules within the *Steuerungszentrum* that have a dictatorial influence, such that quantum-level events that govern their motion, such as Heisenberg's Uncertainty Principle, are amplified to influence the entire organism.

This is an interesting insight and one to which we will return; but it was not developed at the time and it didn't have much influence because, after Germany's defeat in 1945, Jordan's Nazi politics saw him widely discredited among his contemporaries and his ideas in quantum biology neglected. Other matchmakers between the disciplines of biology and quantum physics were scattered to the four winds by the aftermath of the war; and physics, shaken to its core by the use of the atomic bomb, turned its attention to more traditional problems.

But the flame of quantum biology would be kept burning by none other than the inventor of quantum wave mechanics, Erwin Schrödinger. On the eve of the Second World War he fled Austria – his wife was deemed 'non-Aryan' under the Nazi laws – and settled in

Ireland, where in 1944 he published a book whose title posed the question *What Is Life?*, and in which he outlined a novel insight into biology that remains central to the field of quantum biology and indeed to this book. It is this insight that we will explore in a little depth before we end this historical chapter.

Order all the way down

The problem that intrigued Schrödinger was the mysterious process of heredity. You may recall that at this time, in the first half of the twentieth century, scientists knew that genes were inherited from one generation to the next, but not what genes were made of or how they worked. What laws, Schrödinger wondered, provided heredity with its high level of fidelity? In other words, how could identical copies of genes be passed virtually unchanged from generation to the next?

Schrödinger knew that the accurate and repeatedly demonstrable laws of classical physics and chemistry, such as those of thermodynamics, which is driven by the random motion of atoms and molecules, were in reality statistical laws, which means they are only true *on average*, and only reliable because they involve very large numbers of particles interacting. Returning to our billiard table, the motion of a single ball is entirely unpredictable, but throw lots of balls onto the table and knock them randomly about for an hour or so and you can predict that most will have ended up in the pockets. Thermodynamics works like this: it is the average behaviour of lots of molecules that is predictable, not the behaviour of individual molecules. Schrödinger pointed out that statistical laws, such as those of thermodynamics, cease to accurately describe systems composed of just a small number of particles.

Consider, for example, the gas laws described by Robert Boyle and Jacques Charles three hundred years ago. They describe how the volume of gas in a balloon will expand when heated and contract when cooled. This behaviour can be captured in a simple mathematical formula known as the ideal gas law.* A balloon obeys these orderly laws: if you heat it up, it will expand; if you cool it, it will contract. It obeys these laws despite the fact that it is filled with trillions of molecules that behave individually like the disorderly billiard balls whose motion is entirely random, bumping and jostling with each other and bouncing off the inner wall of the balloon. How does disorderly motion generate orderly laws?

When the balloon is heated, the air molecules jiggle about faster, which ensures that they bump into each other, and the walls of the balloon, with a little bit more force. This extra force exerts more pressure on the elastic skin of the balloon (just as it did on the moving baton on Boltzmann's billiard table), causing it to expand. The amount of expansion will depend on how much heat is provided and is entirely predictable and described accurately by the gas laws. The important point is that the singular object that is the balloon strictly obeys the gas law because the orderly motion of its single continuous elastic surface arises from the disorderly motions of very large numbers of particles, generating, as Schrödinger put it, order from disorder.

Schrödinger argued that it is not only the gas laws that derive their accuracy from the statistical properties of large numbers; *all* the laws of classical physics and chemistry – including the laws governing the dynamics of fluids or chemical reactions – are based on this 'averaging of large numbers' or 'order from disorder' principle.

But, although a normal-sized balloon filled with trillions of air molecules will always obey the gas laws, a microscopic balloon, one so tiny it is filled with only a handful of air molecules, will not. This is because, even at constant temperature, this handful of molecules will occasionally and entirely randomly be found moving away from each other, causing the balloon to expand. Similarly, it will occasionally contract for no better reason than that all its molecules randomly

---

* The law is encapsulated in the equation $PV = nRT$, where n is the amount of gas in the sample, R is the universal gas constant, P is the pressure, V is the volume of a gas and T is temperature.

move inwards. The behaviour of a very tiny balloon will thereby be largely unpredictable.

This dependence of orderliness and predictability on large numbers is of course very familiar to us in other walks of life. For example, Americans play more baseball than Canadians, whereas Canadians play more ice hockey than Americans. On the basis of this statistical 'law' one could make additional predictions about each country, for example that America will import more baseballs than Canada and Canada will import more hockey sticks than America. But while such statistical laws have predictive value when applied to whole countries filled with millions of inhabitants, they could not accurately predict trade in hockey sticks or baseballs in a single small town in, say, Minnesota or Saskatchewan.

Schrödinger went further than simply observing that the statistical laws of classical physics could not be relied on at the microscopic level: he quantified the decline in accuracy, calculating that the magnitude of deviations from those laws is inversely proportional to the square root of the number of particles involved. So a balloon filled with a trillion (a million million) particles deviates from the strict behaviour of the gas laws by only one millionth. However, a balloon filled with only a hundred particles will deviate from orderly behaviour by one in ten. Although such a balloon will still tend to expand when heated and contract when cooled, it will not do so in a way that could be captured by any deterministic law. All the statistical laws of classical physics are subject to this restriction: they are true for objects consisting of very large numbers of particles, but they fail to describe the behaviour of objects composed of small numbers of particles. So anything that relies on the classical laws for reliability and regularity needs to be composed of lots of particles.

But what about life? Can its orderly behaviour, such as its laws of heredity, be accounted for by statistical laws? When Schrödinger pondered this question he concluded that the 'order from disorder' principle that underpinned thermodynamics could not govern life – because, as he saw it, at least some of the tiniest biological machines are just too small to be governed by classical laws.

For example, at the time Schrödinger was writing *What Is Life?* – a time when heredity was known to be governed by genes, while the nature of genes was still a mystery – he asked the simple question: are genes big enough to derive their reproductive accuracy from the statistical 'order from disorder' laws? He arrived at an estimated size for a single gene of no bigger than a cube of sides about 300 angstroms (an angstrom is 0.0000001 millimetres). Such a cube would contain about one million atoms. That may sound like a lot, but the square root of a million is a thousand, so the level of inaccuracy or 'noise' in heredity should be of the order of one in a thousand, or 0.1 per cent. So, if heredity were based on classical statistical laws, then it should generate errors (deviations from the laws) at a level of one in a thousand. Yet it was known that genes could be faithfully transmitted with mutation rates (errors) of less than one in one billion. This extraordinary high degree of fidelity convinced Schrödinger that the laws of heredity could not be founded on the 'order from disorder' classical laws. Instead, he proposed that genes were more like individual atoms or molecules in being subject to the non-classical but strangely orderly rules of the science he helped to found, quantum mechanics. Schrödinger proposed that heredity was based on the novel principle of 'order from order'.

He first presented this theory at a series of lectures at Trinity College in Dublin in 1943 and published them the following year in *What is Life?*, in which he wrote: 'The living organism seems to be a macroscopic system which in part of its behaviour approaches to that . . . to which all systems tend, as the temperature approaches the absolute zero and the molecular disorder is removed.' For reasons that we will soon discover, at absolute zero all objects are subject to quantum rather than thermodynamic laws. Life, Schrödinger was claiming, is a quantum-level phenomenon capable of flying in the air, walking on two or four legs, swimming in the ocean, growing in the soil or, indeed, reading this book.

The estrangement

The years following the publication of Schrödinger's book saw the discovery of the DNA double helix and the meteoric rise of molecular biology, a discipline which developed largely without reference to quantum phenomena. Gene cloning, genetic engineering, genome fingerprinting and genome sequencing were developed by biologists who, by and large, were content, with some justification, to ignore the mathematically challenging quantum world. There were occasional forays into the borderland between biology and quantum mechanics. However, most scientists forgot Schrödinger's bold claim; many were even openly hostile to the idea that quantum mechanics was needed to explain life. For example, in 1962 the British chemist and cognitive scientist Christopher Longuet-Higgins wrote:

> I remember some discussion a few years ago about the possible occurrence of long-range quantum-mechanical forces between enzymes and their substrates. It was, however, perfectly right that such a hypothesis should be treated with reserve, not only because of the flimsiness of the experimental evidence but also because of the great difficulty of reconciling such an idea with the general theory of intermolecular forces.[4]

Even in 1993, when the book, *What is Life? The Next Fifty Years* was published,[5] drawing together papers written by participants in a meeting held in Dublin fifty years after Schrödinger's presentation, quantum mechanics was hardly mentioned.

Much of the scepticism Schrödinger's claim attracted at the time was rooted in the general belief that delicate quantum states couldn't possibly survive in the warm, wet and busy molecular environments inside living organisms. As we discovered in the last chapter, this was the principal reason why many scientists were (and many still are) very sceptical towards the notion that the avian compass could be governed by quantum mechanics. You may remember that, when discussing this issue in chapter 1, we described the quantum properties of matter as being 'washed away' by the random arrangement of molecules in big objects. With our thermodynamic insight we can now see the source of that dissipation: it is the billiard-ball-like molecular jostling that Schrödinger identified as the source of the 'order from disorder' statistical laws. Scattered particles can be realigned to reveal their hidden quantum depths, but only in special circumstances and usually only very briefly. For example, we saw how scattered spinning hydrogen nuclei in our body can be lined up to generate a *coherent* MRI signal from the quantum property of spin – but only by applying a very strong magnetic field provided by a big, powerful magnet, and only for as long as that magnetic force is maintained: as soon as the magnetic field is switched off, the particles become randomly aligned again by all the molecular jostling, and the quantum signal becomes scattered and undetectable. This process by which random molecular motion disrupts carefully aligned quantum mechanical systems is known as *decoherence*, and it rapidly wipes out the weird quantum effects in big inanimate objects.

Raising the temperature of a body increases the energy and speed of molecular jostling, so decoherence occurs more readily at higher temperatures. But do not think that 'higher' means hot. In fact, even at room temperature decoherence is almost instantaneous. This is why the idea that warm living bodies could maintain delicate quantum states was, at least initially, considered to be highly implausible. Only when objects are cooled to near absolute zero – a temperature of −273°C – is random molecular motion completely stilled to keep decoherence at bay, allowing quantum mechanics to shine through. The meaning of Schrödinger's statement, quoted above, now becomes clearer. The physicist was claiming that life somehow manages to work to a rulebook that normally operates only at temperatures 273° colder than any living organism.

But, as both Jordan and Schrödinger argued, and as you will discover if you read on, life is different from inanimate objects because relatively small numbers of highly ordered particles, such as those inside a gene or the avian compass, can make a difference to an entire organism. This is what Jordan termed amplification and Schrödinger called order from order. The colour of your eyes, the shape of your nose, aspects of your character, your level of intelligence and even your propensity to disease have in fact all been determined by precisely forty-six highly ordered supermolecules: the DNA chromosomes you inherited from your parents. No inanimate macroscopic object in the known universe has this sensitivity to the detailed structure of matter at its most fundamental level – a level where quantum mechanical rather than classical laws reign. Schrödinger argued that this is what makes life so special. In 2014, seventy years since Schrödinger first published his book, we are finally coming to appreciate the startling implications of the extraordinary answer that he provided to the question: what is life?

3

The engines of life

*Everything that living things do can be understood in terms of the
jiggling and wiggling of atoms . . .*

Richard Feynman[1]

Hamlet: How long will a man lie i' the earth ere he rot?
Gravedigger: Faith, if he be not rotten before he die – as we have
*   many pocky corses now-a-days, that will scarce hold the laying*
*   in, – he will last you some eight year or nine year: a tanner will*
*   last you nine year.*
Hamlet: Why he more than another?
Gravedigger: Why, sir, his hide is so tanned with his trade that he
*   will keep out water a great while; and your water is a sore*
*   decayer of your whoreson dead body.*

William Shakespeare,
Hamlet, Act V, scene i, 'A Churchyard'

SIXTY-EIGHT million years ago, in the period we now call the late
Cretaceous, a young Tyrannosaurus rex was making its way
through a sparsely wooded river valley cut into a semi-tropical forest.
At around eighteen years old, the animal had not yet reached matur-
ity, but it stood nearly 5 metres tall. With every lumbering step it
accelerated many tons of dinosaur meat forward with a momentum
sufficient to flatten trees or any smaller creatures unfortunate enough

to get in its way. That its body could retain its integrity while being subject to these flesh-sundering forces was due to the fact that every bone, sinew and muscle was held in place by tough but elastic fibres of a protein known as *collagen*. This protein acts as a kind of glue that bonds flesh, and it is an essential component of all animal bodies, including our own. Like all biomolecules, it is made and unmade by the most remarkable machines in the known universe. Our focus in this chapter is on how these biological nanomachines* work; and from there we will explore the recent discovery that the gears and levers of these engines of life dip into the quantum world to keep us and every other living organism alive.

But first, back to that ancient valley. On this particular day, the dinosaur's bulk, built by millions of nanomachines, would be its undoing, because those limbs that had been so effective at chasing down and dismembering its prey would prove to be of little use in extricating it from the sticky mud of the soft riverbed into which it stumbled. After many hours of fruitless struggling, the Tyrannosaur's huge jaws filled with murky water and the dying animal sank into the mud. Under most circumstances the animal's flesh would have suffered the same rapid decay as Hamlet's gravedigger's 'corses', but this individual dinosaur sank so fast that its entire body was soon entombed in thick, flesh-preserving mud and sand. Over the years and centuries, finely grained minerals permeated cavities and pores in its bones and flesh, replacing the animal's tissues with stone: the dinosaur corpse became a dinosaur fossil. Up on the surface, the rivers continued to wander over the landscape, depositing successive layers of sand, mud and silt, until the fossil lay beneath tens of metres of sandstone and shale.

About forty million years later the climate warmed, the rivers dried up and the rock layers covering the long-dead bones eroded in hot desert winds. Another twenty-eight million years passed before members of another biped species, *Homo sapiens*, walked into the

---

* 'Nano' refers to structures on the scale of one nanometre, or one-billionth of a metre.

river valley; but these upright primates mostly shunned this dry and hostile country. When, in more modern times, European settlers arrived they named this inhospitable area the Badlands of Montana and called the dry river valley Hell Creek. In 2002, a team of palaeontologists led by the most famous fossil hunter of them all, Jack Horner, was camping there. One of that group, Bob Hormon, was having his lunch when he noticed a large bone jutting out of the rock just above him.

Over the course of three years, nearly half of the entire skeleton of the animal was carefully excavated out of the surrounding stone – a task that involved the Army Corps of Engineers, a helicopter and a lot of graduate students – and transported to the Museum of the Rockies in Bozeman, Montana, where it was designated specimen MOR 1255. The dinosaur's femur had to be cut in two before it could be winched onto a helicopter, and in the process a chunk of fossilized bone was broken off. Jack Horner gave several of the fragments to his palaeontologist colleague, Dr Mary Schweitzer from North Carolina State University, whom he knew to be interested in the chemical make-up of fossils.

When Schweitzer opened the box, she got a surprise. The first fragment she looked at seemed to have very unusual-looking tissue on the inner (marrow-cavity) side of the bone. She placed the bone in an acid bath, which would dissolve its outer stony minerals to reveal its deeper structures. However, on this occasion she accidentally left the fossil in the bath for too long, and by the time she returned *all* of its minerals had dissolved away. Schweitzer expected the entire fossil to have disintegrated, but she and her colleagues were astonished to discover that a pliable fibrous substance remained which, under the microscope, looked just like the kind of soft tissue that you would find in modern bones. And, just as in modern bones, this tissue appeared to be packed full of blood vessels, blood cells and those long chains of collagen fibres, the biological glue that had kept the lumbering live animal in one piece.

Fossils that preserve the structure of soft tissue are rare but far from unknown. The Burgess Shale fossils found high up in the Canadian Rockies of British Columbia between 1910 and 1925 preserve astonishingly detailed impressions of the flesh of animals that swam in the Cambrian seas nearly six hundred million years ago, as does the famous feathered archaeopteryx from the Solnhofen quarry in Germany, which lived some hundred and fifty million years ago. But conventional soft-tissue fossils preserve only the *impression* of biological tissue, not its *substance*; yet the pliable material that remained in Mary Schweitzer's acid bath appeared to be the dinosaur's soft tissue itself. When, in 2007, Schweitzer published her finding in the journal *Science*,[2] her paper was initially met with surprise and a considerable degree of scepticism. But, although the survival of biomolecules for millions of years is indeed astonishing, it is what happened next in this story that is the focus of our interest. To prove that the fibrous structures were indeed made of collagen, Schweitzer first demonstrated that proteins that stick to modern collagen also stuck to the fibres in her ancient bone. As a final test, she mixed the dinosaur tissue with an enzyme called *collagenase,* one of the many biomolecular machines that make and unmake collagen fibres in animal bodies. Within minutes, collagen chains that had held fast for sixty-eight million years were broken by the enzyme.

Enzymes are the engines of life. Those that are probably most familiar to us have somewhat mundane everyday uses, such as the proteases added to 'biological' detergents that help to remove stains, the pectin added to jam to help it set, or the rennet added to milk to help it to coagulate and become cheese. We may also appreciate the role that the various enzymes in our stomachs and intestines play in digesting our food. But these are fairly trivial examples of the action of nature's nanomachines. All life depends or depended on enzymes, from those first microbes that oozed out of the primordial soup, to the dinosaurs that stomped through the Jurassic forests, to every organism alive today. Every cell in your body is filled with hundreds or even thousands of these molecular machines that help to keep that continual process of assembly and recycling of biomolecules, the process that we call life, moving.

Here, 'help' is the key word that defines what enzymes do: their job is to speed up (*catalyse*) all sorts of biochemical reactions that would otherwise proceed far too slowly. Thus, protease enzymes added to detergents speed up the digestion of proteins in stains, pectin enzymes speed up the digestion of polysaccharides in fruit, and rennet enzymes speed up the coagulation of milk. Similarly, enzymes in our cells speed up *metabolism*: the process by which trillions of biomolecules inside our cells are continually transformed into trillions of other biomolecules to keep us alive.

The collagenase enzyme that Mary Schweitzer added to her dinosaur bones is just one of these biomachines whose regular job in animal bodies is to disintegrate collagen fibres. The rate of speed-up provided by enzymes can be roughly estimated by comparing the time taken to digest collagen fibres in their absence (clearly, more than sixty-eight million years) and in the presence of the right enzyme (about thirty minutes): a trillion-fold difference.

In this chapter we'll be exploring how it is that enzymes such as collagenase manage to achieve these astronomical chemical accelerations. One of the surprises of recent years is the discovery that quantum mechanics plays a key role in the action of at least some enzymes; and, since they are central to life, they are our first port of call on the voyage through quantum biology.

Enzymes: between the quick and the dead

The exploitation of enzymes pre-dated their discovery and characterization by many millennia. Several thousand years ago our ancestors were transforming grain or grape juice into beer or wine by the addition of yeast – essentially a microbial bag of enzymes.* They

* Yeasts are single-celled fungi.

also understood that extracts from the stomach lining of calves (rennet) accelerated the transformation of milk into cheese. For many centuries it was believed that these transforming properties were performed by vital forces associated with living organisms, endowing them with the vitality and speed of change that distinguished the living (the 'quick' in the biblical reference in this section heading) from the dead.

In 1752, inspired by the mechanistic philosophy of René Descartes, the French scientist René Antoine Ferchault de Réaumur set out to investigate one of these supposed vital activities, digestion, with an ingenious experiment. It was generally believed at the time that animals digest their food by a mechanical process brought about by pounding and churning within their digestive organs. This theory seemed especially pertinent to birds, whose gizzards contained small stones that were thought to macerate their food – a mechanical action consistent with René Descartes' view (outlined in the previous chapter) that animals were mere machines. But de Réaumur was puzzled by how birds of prey, whose gizzards lacked digestive stones, also managed to digest their food. So he fed his pet falcon small pieces of meat enclosed in tiny metal capsules punctured by small holes. When he recovered the capsules he discovered that the meat was completely digested, despite the fact that, protected within the metal, it could not have been subject to any mechanical action. Descartes' cogs, levers and grinders were clearly insufficient to account for at least one of life's vital forces.

A century after de Réaumur's work, another Frenchman, the chemist and founder of microbiology Louis Pasteur, studied another biological transformation hitherto attributed to 'vital forces': the conversion of grape juice into wine. He showed that the transforming principle of fermentation appeared to be intrinsically associated with living yeast cells that were present in the 'ferments' used in the brewing industry, or in the leaven used to make bread. The term 'enzyme' (Greek: 'in yeast') was then coined by the German physiologist Wilhelm Friedrich Kühne in 1877 to describe the agents of these *vital activities*, such as those performed by living yeast cells, or indeed any transformations promoted by substances extracted from living tissue.

But what are enzymes and how do they quicken life's transformations? Let's return to the enzyme that opened our story in this chapter, collagenase.

Why we need enzymes and how tadpoles lose their tails

Collagen is the most abundant protein in animals (including humans). It acts as a kind of molecular thread woven into and in between our tissues, holding flesh together. Like all proteins, it is composed of basic chemical building blocks: strings of *amino acids* that come in about twenty varieties, of which some (for example, glycine, glutamine, lysine, cysteine, tyrosine) may be familiar to you as nutritional supplements that can be bought in health-food stores. Each amino-acid molecule is made of between ten and fifty or so atoms of carbon, nitrogen, oxygen, hydrogen and occasionally sulphur, held together by chemical bonds in their own uniquely characteristic three-dimensional shape.

Several hundred of these twisted amino-acid molecular shapes are then themselves strung together to form a protein, rather like oddly shaped beads on a string. Each bead is linked to the next via a *peptide bond*, which connects a carbon atom in one amino acid to a nitrogen atom in the next. Peptide bonds are very strong; after all, those that held the T. rex collagen fibres together had survived for sixty-eight million years.

Collagen is an especially strong protein, which is crucial to its role as the internal webbing that maintains the shape and structure of our tissues. The proteins are twisted together in triple strands, which are in turn bonded into thick ropes, or *fibres*. These fibres are threaded through our tissues to sew our cells together; they are also present in tendons, which attach our muscles to our bones, and in ligaments, which bind bone to bone. This dense network of fibres is called the *extracellular matrix* and it basically holds us together.

Anyone who isn't a vegetarian is already familiar with the extracellular matrix as the stringy gristle that you might encounter within an indigestible sausage or in one of the cheaper cuts of meat. Cooks will also be aware of the insolubility of this sinewy material, which fails to tenderize even after hours of boiling a stew. But however unwelcome the extracellular matrix may be on a dinner plate, its presence in the bodies of the diners is absolutely vital. Without collagen, our bones would fall apart, our muscles would drop off our bones and our internal organs would become a kind of jelly.

But the collagen fibres present in your bones, muscles or dinner are not indestructible. Boiling them in strong acids or alkalis will eventually break the peptide bonds between the amino-acid beads and transform these tough fibres into soluble gelatin, the jelly-like substance that is used to make marshmallows and jelly (jello in the US). Film fans might remember the Stay Puft Marshmallow Man in *Ghostbusters* as the giant lumbering mass of wobbly white flesh that terrorized New York. But Marshmallow Man was easily defeated by being liquidized into molten marshmallow cream. Peptide bonds between the amino-acid beads of collagen fibres are the difference between Marshmallow Man and Tyrannosaurus rex. Tough collagen fibres make real animals tough.

There is a problem, however, when you scaffold an animal body with tough, long-lasting materials such as collagen. Consider what happens when you cut or bruise yourself or even break an arm or a leg: tissues are destroyed and the supporting extracellular matrix, that internal stringy mesh, is likely to be damaged or broken. If a house is damaged by a storm or an earthquake, repair has to be preceded by stripping out the broken framework. Similarly, animal bodies use the enzyme collagenase to cut away damaged parts of the extracellular matrix so that the tissue can be repaired – by another set of enzymes.

Even more crucially, the extracellular matrix has to be constantly remodelled as an animal grows: the internal scaffold that sustained an infant will not serve to support the much larger adult. This problem is particularly acute – and its solution therefore particularly instructive – in amphibians, whose adult form is very different from the juvenile. The most familiar example is amphibian metamorphosis: the transformation from a spherical egg to a wriggling tadpole, which later matures into a hopping frog. Fossils of these short-bodied, tail-less amphibians with their unmistakable powerful rear limbs are found in Jurassic rocks dating back to the middle of the Mesozoic Era 200 million years ago, known as the Age of Reptiles. But they can also be found in rocks dating from the Cretaceous period. So it seems likely that frogs swam through that same Montana river where the dinosaur that became MOR 1255 met its end. But, unlike dinosaurs, frogs managed to survive the great Cretaceous extinction and remain common in our own ponds, rivers and swamps, allowing generations of schoolchildren, and scientists, to study how bodies are formed and re-formed.

The transformation of a tadpole into a frog involves a considerable amount of dismantling and reshaping of, for example, the animal's tail, which is gradually reabsorbed into the body and its flesh recycled to form the frog's new limbs. All of this requires the collagen-based extracellular matrix that supported the animal's tail structure to be rapidly dismantled before being reassembled in its newly forming limbs. But, remember those sixty-eight million years under the Montana rocks: collagen fibres are not easily broken. Frog metamorphosis would take a very long time if it relied on the chemical breakdown of collagen solely by inorganic processes. Clearly an animal can't boil its tough sinews in hot acid, and therefore needs a much milder means of dismantling its collagen fibres.

This is where the enzyme *collagenase* comes in.

But how does it – and all its fellow enzymes – work? The vitalist belief that enzyme activity was mediated by some kind of mysterious living force persisted until the late nineteenth century. At that point, one of Kühne's colleagues, the chemist Eduard Buchner, demonstrated that non-living extracts from yeast cells could stimulate precisely the same chemical transformations brought on by the live cells. Buchner went on to make the revolutionary proposal that the *vital force* was nothing more than a form of chemical *catalysis*.

Catalysts are substances that accelerate ordinary chemical reactions and were already familiar to chemists in the nineteenth century. Indeed, many of the chemical processes that drove the industrial revolution depended crucially on catalysts. For example, sulphuric acid was an essential chemical that spurred both the industrial and agricultural revolutions, used in iron and steel manufacture, in the textile industry and for the manufacture of phosphate fertilizer. It is produced by a chemical reaction that starts off with sulphur dioxide (SO_2) and oxygen (the *reactants*), both of which react with water to form the *product*: sulphuric acid (H_2SO_4). However, the reaction is very slow and was therefore initially difficult to commercialize. But in 1831 Peregrine Phillips, a vinegar manufacturer from Bristol, England, discovered a way to speed it up by passing the sulphur dioxide and oxygen over hot platinum, which acted as a *catalyst*. Catalysts differ from the reactants (the initial substances participating in the reaction) because they help to speed up the reaction without taking part in it or being changed by it. Buchner's claim was therefore that enzymes were no different in principle from the kind of inorganic catalyst discovered by Phillips.

Decades of subsequent biochemical research have largely confirmed Buchner's insight. Rennet, produced in calves' stomachs, was the first enzyme to be purified. The ancient Egyptians stored milk in bags made from the lining of calves' stomachs, and it is they who are usually credited with the discovery that this unlikely material accelerated the conversion of milk into the better-preserved cheese. This practice continued until the end of the nineteenth century. By then, calves' stomachs themselves were being dried and sold as 'rennets' in apothecaries' shops. In 1874, the Danish chemist Christian Hansen was being interviewed for a job at an apothecary's when he overheard an order arriving for a dozen rennets and, on enquiring what they were, came up with the idea of using his chemical skills to provide a less unsavoury source of rennet. He returned to his laboratory, where he developed a method for converting the foul-smelling liquid obtained from rehydrating calves' stomachs into a dry powder, and made his fortune by commercializing the product, which was sold the world over as Dr Hansen's Rennet Extract.

Rennet is actually a mixture of several different enzymes, the most active of which for the purposes of cheese-making is called *chymosin*, itself one of a huge family of enzymes called *proteases* that accelerate the cleavage of proteins. Its action in cheese-making is to cause milk to coagulate so it can be separated into curds and whey; but its natural role in a young calf's body is to curdle the milk it ingests so that it remains longer in the digestive tract, giving more time for it to be absorbed. Collagenase is another protease, but methods for its purification weren't developed until fifty years later when Jerome Gross, a clinical scientist at Harvard Medical School in Boston in the 1950s, was intrigued by the question of how tadpoles absorb their tails to become frogs.

Gross was interested in the role of collagen fibres as an example of molecular self-assembly, which he considered to 'hold a major secret of life'.[3] He decided to work on the rather massive tail of the bullfrog tadpole, which can be several inches long. Gross correctly guessed that the process of tail reabsorption must involve a lot of assembly and disassembly of the animal's collagen fibres. To detect collagenase activity he developed a simple test in which a Petri dish was filled with a layer of milky-looking collagen gel, packed full of those tough, durable collagen fibres. When he placed fragments of tissue from tadpole tails on the gel's surface, he noticed a zone surrounding the tissue where those tough fibres were being degraded and turned into soluble gelatin. He then went on to purify the collagen-digesting substance, the enzyme collagenase.

Collagenase is present in the tissue of frogs and other animals, including the dinosaur that left its bones in Hell Creek. The enzyme performed the same function sixty-eight million years ago that it performs today, breaking down collagen fibres; but the enzyme was inactivated when the animal died and fell into the swamp, so its collagen fibres remained intact until Mary Schweitzer added some fresh collagenase to the bone fragments.

Collagenase is just one of millions of enzymes on which all animals, microbes and plants depend to perform nearly all the vital activities of life. Other enzymes make the collagen fibres of the extracellular matrix; yet others make biomolecules such as proteins, DNA, fats and carbohydrates, and a whole different set of enzymes degrade and recycle these biomolecules. Enzymes are responsible for digestion, respiration, photosynthesis and metabolism. They are responsible for making all of us; and they keep us alive. They are the engines of life.

But are enzymes just biological catalysts, providing the same kind of chemistry that is used to make sulphuric acid and scores of other industrial chemicals? A few decades ago, most biologists would have agreed with Buchner's view that the chemistry of life is no different from the kinds of process that take place inside a chemical plant, or even a child's chemistry set. But in the last couple of decades that view has radically changed as a number of key experiments have provided remarkable new insights into the way enzymes work. It seems that life's catalysts are able to reach down into a deeper level of reality than plain old classical chemistry and make use of some neat quantum trickery.

But to understand why quantum mechanics is needed to account for life's vitality, we must first investigate how the far more mundane industrial catalysts work.

Changing the landscape

Catalysts operate by a variety of different mechanisms, but most can be understood through an idea called *transition state theory* (TST)[4] that provides a simple explanation of how catalysts speed up reactions. To understand TST, it is probably useful to first turn the problem around and consider *why* catalysts are needed to accelerate reactions. The answer is that most common chemicals in our environment are rather stable and unreactive. They neither spontaneously break down nor readily react with other chemicals; after all, if they did either of these things, they would not be common now.

The reason why common chemicals are stable is that their bonds are not often broken by the inevitable molecular turbulence that always exists within matter. We can visualize this as the reactant molecules needing to negotiate a landscape, climbing over the top of a hill that stands between them and conversion into products (figure 3.1). The energy needed to ascend the 'hillside' is mostly provided by heat, which speeds the motion of atoms and molecules, causing them to move or vibrate faster. This molecular bumping and jostling can break the chemical bonds that hold the atoms together within molecules and even allow them to form new bonds. But the atoms of more stable molecules – those that are common in our environment – are held together by bonds strong enough to resist the surrounding molecular turbulence. So the chemicals we find around us are

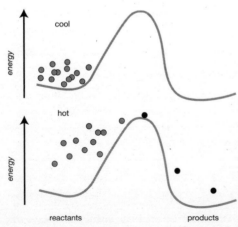

Figure 3.1: Reactant molecules, represented by grey dots, can be converted into product molecules, represented by black dots, but first they have to climb over an energy hill. Cool molecules seldom possess sufficient energy to make the ascent, but hot molecules can easily hike over the summit.

common because their molecules are, by and large, stable,* despite the energetic molecular jostling of their environment.

Even stable molecules can, however, be ripped apart if they are provided with sufficient energy. One possible source of that energy is more heat, which speeds up molecular motion. Heating up a chemical will eventually break its bonds. This is why we cook so much of our food: the heat speeds up the chemical reactions responsible for transforming the raw ingredients – the reactants – into tastier products.

A convenient way to visualize how heat accelerates chemical reactions is to imagine the reactant molecules as the grains of sand in the left-hand chamber of an hourglass lying on its side (figure 3.2a). If left alone all the sand grains will remain where they are until the end of time since they do not possess sufficient energy to reach the neck of the hourglass and pass across to the right chamber, which represents the final products of the reaction. The reactant

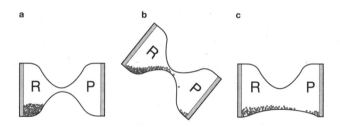

Figure 3.2: Changing the energy landscape. (a) Molecules can pass from the reactant (R) to the product (P) state, but they must first possess sufficient energy to ascend to the transition state (the neck of the hourglass). (b) Turning the hourglass puts the reactant (substrate) into a higher-energy state than the product, allowing it to flow through easily. (c) Enzymes work by stabilizing the transition state, effectively lowering its energy (the neck of the hourglass), allowing substrates to flow through more easily into the product state.

---

*There are, of course, very important exceptions: principally chemicals such as oxygen that, although reactive, are continuously replenished by processes occurring on our planet, most notably those of living organisms such as the plants that pour oxygen into our atmosphere.

molecules in a chemical process can be provided with more energy by heating them up, thereby causing them to move and vibrate faster, and providing some of them with sufficient energy to be converted into products. We can envisage this as simply giving the hourglass a good shake so that some of the sand grains will be thrown into the right-hand chamber and change from reactants to products (figure 3.2b).

But another way of converting reactants to products is to lower the energy barrier they need to climb over. This is what catalysts do. They perform the equivalent of making the neck of the hourglass wider so that sand in the left-hand chamber can flow into the right-hand chamber with only a minimal amount of thermal agitation (figure 3.2c). The reaction is thereby greatly accelerated by the catalyst's ability to change the shape of the energy landscape in such a way as to allow substrates* to become products much faster than they can do in the absence of a catalyst.

We can illustrate how this works at a molecular level by first considering the very slow reaction responsible for breaking down a collagen molecule in the absence of the collagenase[†] enzyme (figure 3.3). As we have already explained, collagen is a string of amino acids, each one attached to the next by a *peptide bond* (shown as a heavy line in the figure) between a carbon and a nitrogen atom. The peptide bond is just one of several types of bond that hold atoms together within molecules. It consists essentially of a pair of electrons that are shared between the nitrogen and carbon atoms. These shared negatively charged electrons attract the positively charged atomic nuclei of the atoms on either side of the bond, thereby acting as a kind of electronic glue that holds the atoms together in the peptide bond.[‡]

* The initial chemical in a reaction is called the reactant; however, when the reaction is helped along by a catalyst such as an enzyme then this initial chemical is referred to as the substrate.
[†] Most enzymes' names start with that of the initial molecule, the 'substrate', that is consumed in the reaction, and end in *-ase*; so collagenase is an enzyme that acts on collagen.
[‡] This type of bond is known as a covalent bond.

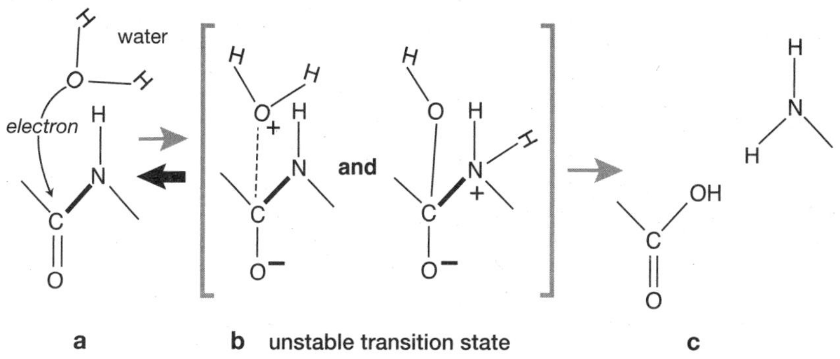

Figure 3.3: Proteins such as collagen (a) consist of chains of amino acids made up from atoms of carbon (C), nitrogen (N), oxygen (O) and hydrogen (H) linked by peptide bonds. One of these bonds is represented by the thick bold line in the figure. The peptide bond can be *hydrolysed* by a water molecule (H_2O), which breaks the peptide bond (c), but it must first pass through an unstable transition state, which consists of at least two different structures that mutually interconvert (b).

Peptide bonds are very stable because breaking them, by forcing the shared electrons to separate, requires a high 'activation energy': the bond has to climb a very tall energy hill before it reaches the neck of the reaction hourglass. In practice, the bond doesn't usually break of its own accord and needs a helping hand from one of the surrounding water molecules in a process known as *hydrolysis*. For this to take place, the water molecule must first wander close enough to the peptide bond to donate one of its electrons to the bond's carbon atom, forming a new weak bond that tethers the water molecule in place, represented by dotted lines in figure 3.3. This intermediate stage is called a transition state (hence *transition state theory*) and is the unstable peak of the energy hill that needs to be climbed if the bond is to be broken, represented by the neck of the hourglass. Note from the figure that this donated electron from water has travelled all the way down to the oxygen atom adjacent to the peptide bond, which having acquired an extra electron is now negatively charged. The water molecule that donated the electron meanwhile has been left with an overall positive charge in the transition state.

Here is where the process gets slightly trickier to grasp. Think of this water molecule (H_2O) as positively charged not because it has lost an electron, but because it now contains a bare hydrogen nucleus, a proton, represented by the + sign in the figure. This positively charged proton is no longer held firmly in place within the water molecule and becomes *delocalized* in the quantum mechanical sense we discussed in the last chapter. Although it spends most of its time still associated with its water molecule (the left-hand structure in figure 3.3b), some of the time it can be found further away, closer to the nitrogen atom (the right-hand structure in the centre of figure 3.3b) at the other end of the peptide bond. In this position, the roving proton can tug one of the peptide bond electrons out of its position, thereby breaking the bond.

But this will not usually happen. The reason is that transition states, such as the one illustrated in figure 3.3b, are very short-lived; they are so unstable that the slightest 'nudge' can dislodge them. For example, the negatively charged electron that was donated by the water molecule is easily reclaimed so that the initial reactants are re-formed (shown by the thick arrow in the figure). This is a far more likely scenario than the forward reaction in which the bond gets broken. So peptide bonds usually don't break. In fact, in neutral solutions, which are neither acidic nor alkaline, the time taken for half the peptide bonds in a protein to break, known as the half-life of the reaction, is more than 500 years.

All this, of course, is what happens *without* enzymes: we have yet to describe how the enzyme comes in to help the hydrolysis process. According to transition state theory, catalysts speed up chemical processes, such as the breaking of the peptide bond, by making the transition state more stable, thereby increasing the chances of the final products forming. There are various ways this can happen. For example, a positively charged metal atom near the bond can neutralize the negatively charged oxygen atom in the transition state to stabilize it (so that it is no longer in such a hurry to give back the electron donated by the water molecule). By stabilizing transition states, catalysts are lending a helping hand by performing the equivalent of widening the neck of the hourglass.

We now need to consider whether transition state theory, viewed through our hourglass analogy, can also account for the way enzymes accelerate all those other reactions necessary for life.

Jiggling and wiggling

The collagenase enzyme that Mary Schweitzer used to shatter those ancient Tyrannosaurus collagen fibres is the same enzyme that Jerome Gross detected in frogs. You will remember that this enzyme is needed to dismantle the tadpole's extracellular matrix so that its tissues, cells and biomolecules can be reassembled into an adult frog. It performed the same function in the dinosaur, and continues to perform that function in our bodies: dismantling collagen fibres to allow growth and re-formation of tissue during development and after injury. To see this enzymatic process in action, we will borrow an idea from a science-transforming lecture delivered by Richard Feynman to an audience at the California Institute of Technology in 1959 entitled 'There's Plenty of Room at the Bottom'. The lecture is generally acknowledged as having been the intellectual foundation of the field of nanotechnology: engineering on the scale of atoms and molecules. Feynman's ideas are also said to have inspired the 1966 film *Fantastic Voyage*, in which a submarine and its crew were shrunk small enough to be injected into a scientist's body to find and repair a potentially fatal blood clot in his brain. To investigate how it works we will take a trip in an imaginary nanosubmarine. Our destination will be the tail of a tadpole.

First we must find our tadpole. A visit to the local pond reveals a clutch of frogspawn, and we carefully remove a handful of the jelly-like black-dotted spheres and transfer them to a glass water tank. It isn't long before we observe wriggling within the spawn and, within days, tiny tadpoles emerging from their eggs. After making a quick note of their principal features under a magnifying glass – a relatively large head with snout above a small mouth, lateral eyes, and feathery gills in front of a long powerful tail fin – we supply the tadpoles with sufficient food (algae) and return daily for observations. For several weeks we notice little change in the form of the animal but are impressed by its rapid increase in length and girth. By about eight weeks we notice that the animal's gills have retracted into its body, revealing front limbs. Another two weeks and rear legs emerge from the base of its sturdy tail. At this stage we must make more frequent observations since the rate of change in the metamorphosing animal appears to be accelerating. The tadpole's gills and gill pouches completely disappear and its eyes migrate higher up its head. Alongside these dramatic changes to the tadpole's front end, its tail starts to shrink. This is the cue we have been waiting for: so we board our nanosubmarine and launch it into the glass tank to investigate one of nature's most remarkable transformations.

As our craft shrinks we can see more detail of the frog's metamorphosis, including dramatic changes to the tadpole's skin, which has become thicker, tougher and embedded with mucus-secreting glands that will keep it moist and supple when it leaves the pond and walks onto land. We dive into one of these glands, which leads us through the animal's skin. After safely passing through several cell barriers, we arrive within its circulatory system. Cruising through the animal's veins and arteries, we can witness from the inside the many changes taking place within its body. From their sac-like beginnings, its lungs form, expand and fill with air. The tadpole's long spiral gut, which was suitable for digesting algae, is straightened into one typical of a predator. Its translucent cartilaginous skeleton, including the notochord (a primitive form of backbone that runs the length of its body), becomes dense and opaque as cartilage is replaced by bone. Continuing our mission, we follow the developing spine down into the tail of the tadpole, which is just beginning the process of being absorbed into the growing body of the frog. At this scale we can see thick striated muscle fibres packed into its length.

Another round of shrinkage allows us to see that each muscle fibre is composed of long columns of cylindrical cells whose periodic contractions are the source of the tadpole's locomotion. Surrounding these muscle cylinders is a dense netting of stringy ropes: the extracellular matrix that is the target of our investigation. The matrix itself appears to be in a state of flux as individual ropes are unravelling to release trapped muscle cells that break free to join a growing mass cell migration out of the disappearing tadpole tail and into the frog's body.

Shrinking down further, we home in on one of those unravelling ropes of the disintegrating extracellular matrix. As its girth expands we see that, like a rope, it is woven from thousands of individual protein cords, each of which is itself a bundle of collagen fibres. Each fibre is made of three collagen protein strings – those amino-acid beads on a string that we met earlier when discussing the dinosaur bone – but wound around each other to make a tough helical thread, a bit like DNA but triple-stranded rather than double-stranded. And here at last we spot the target of our expedition: a collagenase enzyme molecule. It shows up as a clam-like structure clamped onto one of the collagen fibres, and slides down the fibre, unzipping the triple helix strands before simply clipping apart the peptide bonds connecting the amino-acid beads. The chain that might otherwise remain intact for millions of years is broken in an instant. We will now zoom in even further to see exactly how this clipping action works.

Our next bout of shrinkage takes us down to the molecular scale of just a few nanometres (millionths of a millimetre). It is difficult to grasp just how minute this scale actually is, so to give you a better idea, consider the size of the letter 'o' on this page: if you were to shrink down from your normal size to the nanometre scale, then to you that 'o' would appear to be roughly the size of the whole of the United States of America. At this scale we can see that the interior of the cell is densely packed with water molecules, metal ions* and

---

* An ion is an atom or molecule that carries an electrical charge as a result of having missing electrons (positive ion) or additional electrons (negative ion).

a vast and diverse variety of biomolecules that include lots of those oddly-shaped amino acids. This busy and crowded molecular pond is in a state of constant agitation and turbulence, with the molecules spinning and vibrating and bouncing off each other in that billiard-ball-like molecular motion that we met in the last chapter.

And there, among all this randomly turbulent molecular activity, are those clam-like enzymes sliding along the collagen fibres, moving in a very different way. At this scale we can zoom in on a single enzyme as it clips its way along the collagen protein chain. At first sight, the overall form of the enzyme molecule looks rather lumpy and amorphous, giving the false impression that it is a rather disorganized assembly of parts. But collagenase, like all enzymes, has a precise structure, with every atom occupying a specific location within the molecule. And, in contrast to the random molecular jostling of the surrounding molecules, the enzyme is performing an elegant and precise molecular dance as it wraps itself around the collagen fibre, unwinding the fibre's helical turns and precisely snipping the peptide bonds that link the amino acids in the chain before unwrapping itself and moving along to clip the next peptide bond in the chain. These are not shrunken-down versions of manmade machines whose operations are, at a molecular level, driven by the chaotic billiard-ball-like motion of trillions of randomly moving particles. These nanomachines of nature are performing, at a molecular level, a carefully choreographed dance whose actions have been precision engineered by millions of years of natural selection to manipulate the motion of the fundamental particles of matter.

To get a closer look at the cutting action, we descend into the enzyme's jaw-like cleft that holds the substrates in place: the collagen protein chain and a single water molecule. This is the *active site* of the enzyme – its business end that is speeding up the breaking of peptide bonds by bending the neck of the energy hourglass. The choreographed action taking place within this molecular steering centre is very different from all the random jostling going on outside and around the enzyme, and it plays a disproportionately important role in the life of the entire frog.

The enzyme's active site is illustrated in figure 3.4. By comparing this diagram with figure 3.3, you can see that the enzyme is restraining the peptide bond in the unstable transition state that has to be reached before the bond can be broken. The substrates are tethered by weak chemical bonds, indicated by dotted lines in the figure, which are essentially electrons that are shared between the substrate and the enzyme. This tethering holds the substrates in a precise configuration ready for the chopping action of the enzyme's molecular jaws.

As the jaws of the enzyme close, they do something far subtler

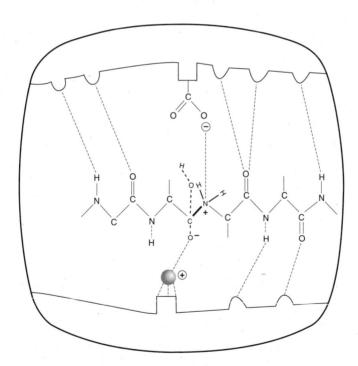

Figure 3.4: Breakage of the peptide bond (in bold) of collagen at the active site of collagenase. The transition state of the substrate is shown by the dashed lines. The sphere to the left of centre at the bottom is the positively charged zinc ion; the top carboxyl group (COO) is from a glutamate amino acid at the enzyme active site. Note that molecular distances are not drawn to scale.

than simply 'biting down' on the bond: they provide the means through which catalysis can take place. We notice a big positively charged atom hanging directly beneath the target peptide bond being swung into position. This is a positively charged zinc atom. If we consider the active site of the enzyme to be its jaws, then the zinc atom is one of its two incisors. The positively charged atom plucks an electron out of the oxygen atom from the substrates to stabilize the transition state and thereby deform the energy landscape: the hourglass has just had its neck widened.

The rest of the job is carried out by the enzyme's second molecular incisor. This is one of the enzyme's own amino acids called glutamate, which has swung into position to hang its negatively charged oxygen atom over the target peptide bond. Its role is first to pluck a positively charged proton out of the tethered water molecule. It then spits this proton into the nitrogen atom at one end of the target peptide bond, giving it a positive charge which draws electrons out of the peptide bond. You may remember that electrons provide the glue of chemical bonds; so drawing the electron out is like pulling the glue out of a bonded joint, causing it to weaken and break.[5] A few more electron rearrangements and the products of the reaction, the broken peptide chains, are expelled from the enzyme's molecular jaws. A reaction that might otherwise take upwards of sixty-eight million years has been completed in nanoseconds.

But where does quantum mechanics come into the picture? To appreciate why we need quantum mechanics to explain enzyme catalysis, we will pause for a moment to consider again those insights provided by the quantum mechanics pioneers. We have already mentioned the special role played by those few particles at the active site of the enzyme whose choreographed motions are in stark contrast to the random molecular jostling going on elsewhere in the molecular environment. Here, highly structured biomolecules interact in very specific ways with other highly structured biomolecules. This can be seen as either Jordan's dictatorial amplification or Erwin Schrödinger's 'order from order' that goes all the way down from the developing frog through its organized tissues and cells down to the fibres that hold those tissues and cells together and the choreographed motion of fundamental particles within the active site of collagenase that remodels those fibres and thereby affects the development of the entire frog. Whether we choose Jordan's model or Schrödinger's, what is going on here is clearly very different from the chaotic molecular motion that pushes trains up hillsides.

But does this molecular order allow a different set of rules to come into play in life, as Schrödinger claimed? To discover the answer to this question we need to know a little more about that different set of rules that operates at the scale of the very small.

Does transition state theory explain it all?

Does such choreographed molecular motion necessarily involve quantum mechanics? We have discovered that the ability of collagenase to accelerate the breakage of peptide bonds involves several of the catalytic mechanisms that chemists routinely use to accelerate chemical reactions, without recourse to quantum mechanics. For example, the zinc metal atom at the active site of the enzyme appears to be playing a similar role to the hot platinum metal that Peregrine Phillips used in the nineteenth century to accelerate the manufacture of sulphuric acid. These inorganic catalysts rely on random molecular motions, rather than choreographed actions, to bring their catalytic groups close to their substrates and thereby accelerate their chemical reactions. Is enzyme catalysis just a collection of several straightforward classical catalytic mechanisms packed into active sites, thereby providing the vital spark that ignites life?

Up until recently, nearly all enzymologists would have said yes; standard transition state theory, with its description of the different processes that help extend the life of the intermediate transition state, was considered to be the best explanation of how enzymes work. But after all the known contributing factors were taken into account, some doubts emerged. For example, the different possible mechanisms that can speed up the peptide cleavage reaction discussed earlier in this chapter are each well understood and give rise, individually, to rate enhancement factors of up to about a hundredfold. But even if you multiply all these factors together, the most that can be achieved is about a million-fold enhancement in reaction rate. This is a puny number compared to the kinds of rate enhancement that enzymes are known to deliver: there seems to be an embarrassingly large gap between theory and reality.

Another puzzle is how enzyme activity is affected by various kinds of change to the structure of the enzymes themselves. For example, like all enzymes, collagenase consists essentially of a protein chassis on a string that supports the jaws and teeth of the enzyme within its active site. We would expect that changing the amino acids that form its jaws and teeth would have a big impact on an enzyme's efficiency, and indeed it does. What is more surprising is the discovery that changing amino acids within the enzyme that are far from the active site can also have dramatic effects on its efficiency. Why these supposedly innocuous modifications to enzyme structure make such a dramatic difference remains something of a mystery within standard transition state theory; but it turns out that they make sense if quantum mechanics is brought into the picture. We will return to this discovery in the last chapter of the book.

Yet another problem is that transition state theory has so far failed to deliver artificial enzymes that work as well as the real ones. You may remember Richard Feynman's famous dictum, 'What I cannot create, I do not understand.' This is relevant to enzymes because, despite knowing so much about enzyme mechanisms, no one has so far managed to design an enzyme from scratch that can produce anything like the rate enhancements delivered by natural enzymes.[6] According to Feynman's criterion, we do not yet understand how enzymes work.

But take another look at figure 3.4 and ask the question: what is the enzyme doing? The answer is pretty obvious: enzymes manipulate individual atoms, protons and electrons, within and between molecules. Up to now in this chapter we have considered these particles as behaving pretty much as though they were tiny lumps of electric charge being pushed and pulled from one place to another in ball-and-stick-like molecules. But as we saw in our explorations of the last chapter, electrons, protons and even whole atoms are very different from such classical balls because they adhere to the rules of quantum mechanics, including the weird ones that depend on coherence but are normally filtered out at the macroscopic level of billiard balls by that process of decoherence. Billiard balls are not, after all, good models for fundamental particles; so, to understand the real action that goes on inside the active sites of enzymes, we must leave our classical preconceptions behind and enter the weird world of quantum mechanics where objects can be doing two or a hundred things at once, can possess spooky connections and can pass through apparently impenetrable barriers. These are feats that no billiard ball has ever accomplished.

Pushing electrons around

As we have discovered, one of the key activities of enzymes is to move electrons around within substrate molecules, as for example when collagenase pushes and pulls electrons within the peptide molecule. But as well as being pushed around within molecules, electrons can also be transferred from one molecule to another.

A very common type of electron transfer reaction in chemistry takes place during a process called *oxidation*. This is what happens when we burn carbon-based fuels, such as coal, in air. The essence of oxidation is movement of electrons from a donor to an acceptor molecule. In the case of burning a lump of coal, high-energy electrons from carbon atoms move to form lower-energy bonds within oxygen atoms, giving rise to carbon dioxide. The surplus energy is released as the heat of a coal fire. We harness this thermal energy to heat our homes, cook our food and turn water into the steam that drives an engine or powers a turbine to generate electricity. But the burning of coal and internal combustion engines are fairly crude and inefficient devices for utilizing electron energy. Nature long ago discovered a far more efficient means of capturing this energy, through the process of respiration.

We tend to think of respiration as the process of breathing: taking the oxygen we need into our lungs and expelling carbon dioxide as a waste product. But breathing is in fact a combination of just the first (the delivery of oxygen) and last (the expulsion of carbon dioxide) steps of a far more complex and orderly molecular process that goes on within all our cells. It takes place inside complex organelles* called *mitochondria*, which look a bit like bacterial cells trapped inside our own larger animal cells, since they too have internal structures such as membranes and even their own DNA. In fact, mitochondria almost certainly evolved from a symbiotic bacterium that made a home inside the ancestor of animal and plant cells hundreds of millions of years ago and then lost the ability to live independently. But their ancestry as an independently living bacterial cell probably explains why they are capable of executing such an extraordinarily intricate process as respiration. In fact, in terms of chemical complexity, respiration is probably second only to photosynthesis, which we will meet in the next chapter.

To home in on the role that quantum mechanics plays here, we will need to simplify how respiration works. And even when simplified, it still involves a remarkable sequence of processes that beautifully convey the wonder of these biological nanomachines. It starts off with the burning of a carbon-based fuel, in this case the nutrients we get from our food. For example, carbohydrates are broken down in our gut to yield sugars, such as glucose, that are loaded into the bloodstream and then delivered to cells hungry for energy. The oxygen needed to burn this sugar fuel is delivered by the blood from the lungs to the same cells. Just as with the burning of coal, electrons in the outer orbits of carbon atoms within a molecule

---

* As you may remember from chapter 2, organelles are the 'organs' of cells: internal structures that perform particular functions, such as respiration.

are transferred to a molecule called NADH. But instead of being used immediately to bond to the oxygen atoms, the electrons are passed from one enzyme to another along a *respiratory chain* of enzymes inside our cells, rather like the baton being passed from one runner to another in a relay race. At each transfer step the electron is dropped into a lower-energy state and the difference in energy is used to power enzymes that pump protons out of the mitochondria. The resulting proton gradient from the outside to the inside of the mitochondria is then used to drive the rotation of another enzyme, called ATPase, which makes a biomolecule called ATP. ATP is very important in all living cells as it acts as a kind of energy battery that can easily be transported around the cell to power lots of energy-hungry activities, such as moving or building bodies.

The function of the electron-driven proton-pumping enzymes is a bit like that of hydroelectric pumps that store excess energy by pumping water up a hillside. The stored energy can then be released by letting the water flow down the hillside to rotate a turbine engine that generates electrical power. Similarly, respiratory enzymes pump protons out of the mitochondria. When the protons flow back inside, they power the rotations of the turbine-like ATPase enzyme. These rotations drive another set of choreographed molecular motions that bolt a high-energy chemical phosphate group onto a molecule within the enzyme to make ATP.

Extending the analogy of this energy-capturing process as a relay race, we can imagine the baton being replaced by a bottle of water (representing the electron energy), with each runner (enzyme) taking a sip of water and then passing on the bottle, before finally the remainder of the water is poured into a bucket called oxygen. This capturing of the electron energy in small chunks makes the whole process much more efficient than simply pouring it directly into oxygen, as very little of it is lost as waste heat.

So the key events of respiration actually have very little to do with the process of breathing, but consist instead of an orderly transfer of electrons through a relay of respiratory enzymes inside our cells. Each electron transfer event, between one enzyme and the next in the relay, takes place across a gap of several tens of angstroms – a distance of many atoms – far further than was thought to be possible for conventional electron-hopping. The puzzle of respiration is how these enzymes are able to shift the electrons so quickly and efficiently across such big molecular gaps.

This question was first asked as far back as the early 1940s by the Austro-Hungarian–American biochemist Albert Szent-Györgyi, who won the Nobel Prize in medicine in 1937 for his part in the discovery of vitamin C. In 1941, Szent-Györgyi delivered a lecture entitled 'Towards a New Biochemistry' in which he proposed that the way electrons flow easily through biomolecules is similar to how they move in semiconductor materials such as the silicon crystals used in electronics. Unfortunately, it was realized just a few years later that proteins are in fact rather poor conductors of electricity, so electrons would not easily flow through the enzymes in the way that Szent-Györgyi envisaged.

Major advances in chemistry were made during the 1950s, in particular by the Canadian chemist Rudolph Marcus who developed a powerful theory that is today named after him (Marcus theory) and which explains the rate at which electrons can move or jump between different atoms or molecules. He too eventually received a Nobel Prize, in chemistry, in 1992 for his work.

But half a century ago, the issue of how respiratory enzymes in particular were able to encourage such rapid transfer of electrons across relatively large molecular distances remained a puzzle. One suggestion was that proteins might rotate in sequence like clockwork machines, bringing distant molecules close together so that the electrons could easily hop across. An important prediction of these models was that this mechanism would slow down dramatically at low temperatures, when there is less thermal energy to drive the clockwork motion. Yet in 1966, one of the very first real breakthroughs in quantum biology came from experiments carried out at the University of Pennsylvania by two American chemists, Don

DeVault and Britton Chance, who showed that, contrary to all expectations, the rate of electron-hopping in respiratory enzymes did not drop at low temperatures.[7]

Don DeVault was born in Michigan in 1915 but moved west with his family during the Depression. He studied at Caltech and Berkeley in California and received a PhD in chemistry in 1940. He was a committed human rights activist and spent time in prison during the Second World War for his stance as a conscientious objector. In 1958, he resigned his post as professor of chemistry at the University of California to move to Georgia so that he could be directly involved in the struggle for racial equality and integration in the South. His strength of conviction, dedication to the cause and adherence to peaceful protest exposed him to the risk of physical attack during marches with black activists. He even had his jaw broken on one occasion when his racially mixed group of protesters were attacked by a mob. But this didn't deter him.

In 1963, DeVault went to work at the University of Pennsylvania with Britton Chance, a man just two years his senior who had already established a worldwide reputation as one of the leading scientists in his field. Chance had obtained not one but two PhDs, the first in physical chemistry and the other in biology. So his 'field' of expertise was very wide and his research interests diverse. He had spent much of his career working on the structure and function of enzymes – while taking time out to win a gold medal in sailing for the United States in the 1952 Olympics.

Britton Chance had been intrigued by a mechanism by which light can promote the transfer of electrons from the respiratory enzyme cytochrome to oxygen. Together with Mitsuo Nishimura, Chance found that this transfer takes place in the bacterium *Chromatium vinosum* even when its cells are cooled to a chilly liquid nitrogen temperature of −190°C.* But how this process varied with

---

* Most scientists use the Kelvin (K) as their unit of temperature. A 1 K change in temperature corresponds to a 1°C change. However, the Kelvin scale starts at what is called absolute zero, which is equivalent to −273°C. So, for example, human body temperature is 310 K.

changing temperature, which might provide clues to the molecular mechanism involved, was still unknown. What was required, Chance realized, was to initiate the reaction very rapidly with a very brief, but intense, flash of light. This is where Don DeVault's expertise came in. He had spent some years working as an electrical consultant for a small company developing a laser that could provide just such short light pulses.

Together, DeVault and Chance designed an experiment in which a ruby laser delivered a brief flash of bright red light for just 30 nanoseconds (30 billionths of a second) to bacterial cells packed full of respiratory enzymes. They found that as they reduced the temperature the rate of electron transfer fell until, at about 100 K (or −173°C), the electron transfer reaction time was about a thousand times slower than it was at room temperature. This was to be expected if the process of electron transfer was driven primarily by the amount of thermal energy involved. However, something odd happened when DeVault and Chance reduced the temperature below 100 K. Instead of dropping to lower values, the rate of electron transfer seemed to have reached a plateau, remaining constant despite further reduction in temperature, right down to 35 degrees above absolute zero (−238°C). This indicated that the electron transfer mechanism cannot be due solely to the 'classical' electron-hopping described earlier. The answer, it seems, lies in the quantum world, specifically in the weird process of quantum tunnelling that we met in chapter 1.

Quantum tunnelling

You may remember from chapter 1 that quantum tunnelling is the peculiar quantum process that allows particles to pass through impenetrable barriers as easily as sound passes through walls. It was first discovered in 1926 by the German physicist Friedrich Hund and was soon after used successfully to explain the concept of radioactive decay by George Gamow, Ronald Gurney and Edward Condon, all using the then new mathematics of quantum mechanics. Quantum tunnelling became a staple feature of nuclear physics, but it was later appreciated as a phenomenon that applied more widely in material science and chemistry. As we have already seen, it is essential for life on earth as it allows pairs of positively charged hydrogen nuclei in the interior of the sun to fuse together in the first step of converting hydrogen to helium, thereby releasing the sun's vast energy. However, until recently, it was not thought to be involved in any living processes.

One way of thinking about quantum tunnelling is as a means by which particles can get from one side of a barrier to the other in a way that common sense tells us should be impossible. By 'barrier' we mean here a physically impassable region of space (without sufficient energy) – think of force fields used in science fiction stories. This region could consist of a narrow insulating material separating two sides of electric conductors or even empty space, such as the gap between two enzymes in a respiratory chain. It can also be the kind of energy hill that we described earlier, which limits the rate of chemical reactions (figure 3.1). Consider the example of a ball being kicked up a small hill. In order for it to reach the top and roll down the other side it has to be given a firm enough kick. As it climbs the slope it will gradually slow down, and without sufficient energy (a hard enough kick) it will simply stop and roll back again the way it came. According to classical Newtonian mechanics, the only way a ball can get across the barrier is for it to possess sufficient energy to be lifted over the energy hill. But if that ball were an electron, say, and the hill a repulsive energy barrier, then there would be a small probability that the electron would flow through the barrier as a wave, essentially making an alternative and more efficient passage through. This is quantum tunnelling (figure 3.5).

An important feature of quantum mechanics is that the lighter the particle, the easier it is for it to tunnel. It is not surprising, therefore, that once this process was understood to be a ubiquitous feature of the subatomic world it was the tunnelling of electrons that was

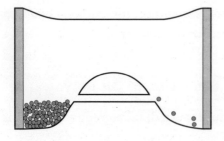

Figure 3.5: Quantum tunnelling through the energy landscape.

found to be most common as they are very light elementary particles. The field emission of electrons from metals was explained as a tunnelling effect in the late 1920s. Quantum tunnelling also explained how radioactive decay takes place: when certain atomic nuclei such as those of uranium occasionally spit out a particle. This became the first successful application of quantum mechanics to the problems of nuclear physics. In chemistry, quantum tunnelling of electrons, protons (hydrogen nuclei) and even heavier atoms is today well understood.

A crucial feature of quantum tunnelling is that, like many other quantum phenomena, it depends on the spread-out wave-like nature of matter particles. But for a body made up of very many particles to tunnel it has to maintain the wave aspects of all its constituents marching in step, with peaks and troughs of waves coinciding, something we refer to as the system being coherent, or simply 'in tune'. Decoherence describes the process whereby all the many quantum waves very rapidly get out of step with each other and wash away any overall coherent behaviour, thus destroying the body's ability to quantum tunnel. For a particle to quantum tunnel, it must remain wavy in order to seep through the barrier. This is why big objects, such as footballs, do not quantum tunnel: they are made up of trillions of atoms that cannot behave in a coordinated coherent wave-like fashion.

By quantum standards, living cells are also big objects, so at first glance it would seem unlikely that quantum tunnelling would be found inside hot, wet living cells whose atoms and molecules would mostly be moving incoherently. But, as we have discovered, the interior of an enzyme is different: its particles are engaged in a choreographed dance rather than a chaotic rave. So let us explore how this choreography can make a difference to life.

Quantum tunnelling of electrons in biology

It took several years for the unexpected temperature profile of DeVault and Chance's 1966 experiment to be fully explained. Another American scientist whose work spanned many disciplines, ranging from molecular biology to physics to computer science, is John Hopfield. Best known for his work on developing neural networks in computing, Hopfield was nevertheless very interested in the physical processes involved in biology. In 1974 he published a paper entitled 'Electron transfer between biological molecules by thermally activated tunneling',[8] in which he developed a theoretical model to explain the DeVault and Chance result. Hopfield pointed out that at high temperature the vibrational energy of the molecules would be sufficient to allow the electrons to hop over the top of a barrier without tunnelling. As the temperature is reduced, there shouldn't be enough vibrational energy for the enzymatic reaction to take place. But DeVault and Chance had found that the reaction did proceed at low temperatures. Hopfield therefore suggested that at these lower temperatures the electron is raised to a state sitting halfway up the energy slope, where the distance it needs to traverse is shorter than it is at the bottom of the slope, enhancing its chances of quantum tunnelling through the barrier. And he was right: the tunnelling-mediated transfer of electrons takes place even at very low temperatures, just as DeVault and Chance found.

Few scientists now doubt that electrons travel along respiratory chains via quantum tunnelling. This places the most important energy-harnessing reactions in animal and (non-photosynthetic) microbial cells (we will be dealing with the photosynthetic sort in the next chapter) firmly within the sphere of quantum biology. But electrons are very light, even by the standards of the quantum world, and their behaviour is inevitably very 'wave-like'. They should not therefore be regarded as moving and bouncing about like tiny classical particles, despite the fact that they are still treated this way in many standard biochemistry texts that continue to use the 'solar system' model of the atom. A much more appropriate representation of the electrons in an atom is as a spread-out, wavy cloud of 'electronness' surrounding the tiny nucleus, the 'cloud of probability' that we discussed in chapter 1. It is perhaps not so surprising, therefore, that electron waves can pass through energy barriers rather like sound waves passing through walls, as we described in that first chapter, even in biological systems.

But what about bigger particles, such as protons or even whole atoms? Can these also tunnel in biological systems? At first glance you would think the answer would be no. Even a single proton is two thousand times as heavy as an electron, and quantum tunnelling is known to be exquisitely sensitive to how massive the tunnelling particle is: small particles tunnel readily whereas heavy particles are far more resistant to tunnelling unless the distances to be covered are very short. But recent remarkable experiments indicate that even these relatively massive particles are able to quantum tunnel in enzymatic reactions.

Moving protons around

You may remember that, as well as promoting electron transfer, one of the key activities of the enzyme collagenase (figure 3.4) is moving protons to promote breaking of the collagen chain. As already mentioned, this kind of reaction is one of the most common particle manipulation tricks performed by enzymes. About one-third of all enzyme reactions involve moving a hydrogen atom from one place to another. Note here that 'hydrogen atom' can mean several things: it could be a neutral atom of hydrogen (H) consisting of an electron around its nucleus (a proton); it could be a positively charged hydrogen ion (H^+), which is just a bare nucleus – a proton without its electron; or it could even be a negatively charged hydride ion, which is a hydrogen atom with an extra electron (H^-).

As any self-respecting chemist or biochemist will quickly tell you, moving hydrogen atoms (well, protons) around within and between molecules does not necessarily imply any quantum effect; or at least, none that requires us to appeal explicitly to the weirder processes of the quantum world, such as tunnelling. Indeed, for most chemical reactions occurring at the kind of temperatures at which life operates, protons are thought to move mostly by non-quantum thermal hopping from one molecule to another. But proton tunnelling is involved in a few chemical reactions that can be identified by their relative indifference to temperature, just as DeVault and Chance had demonstrated for electron tunnelling.

Life operates at high temperatures (by the standards of the quantum world). So, for most of the history of biochemistry, scientists assumed that enzymatic transfer of protons was mediated entirely by the (non-quantum) mechanism of hopping over the energy barrier.* But this view changed in 1989 when Judith Klinman and her colleagues at Berkeley provided the first direct evidence for proton tunnelling in enzyme reactions.[9] Klinman is a biochemist who has long argued for the importance of proton tunnelling in the molecular machinery of life. Indeed, she has gone so far as to claim that it is one of the most important and prevalent mechanisms in the whole of biology. Her breakthrough came from a study of a particular enzyme in yeast called alcohol dehydrogenase (ADH), whose job it is to transfer a proton from an alcohol molecule to another small molecule called NAD^+ to form NADH (nicotinamide adenine dinucleotide, a molecule we have met already as the cell's principal electron carrier). The team were able to confirm the presence of

---

*You might wonder why quantum tunnelling is therefore necessary to explain the fusion processes inside the sun. But there, even the incredibly high temperatures and pressures are not enough to overcome the electrical repulsion that prevents the fusion of the two positively charged protons, and so quantum mechanics is needed to offer a helping hand.

proton tunnelling by using an ingenious technique called the *kinetic isotope effect*. This idea is well known in chemistry and deserves a careful explanation here, for it helps provide one of the main pieces of evidence for quantum biology and will crop up a few times more in this book.

The kinetic isotope effect

Have you ever cycled up a steep hill only to find yourself being overtaken by people on foot? On level ground, you have no problem cycling effortlessly past any number of pedestrians, even runners, so why is cycling so much less efficient on hills?

Imagine that instead of cycling you got off the saddle and walked the bicycle either along the flat ground or up the hill. Now, the issue becomes obvious. On the hill, you have to push the bicycle as well as yourself up the incline. The weight of the bicycle, which was pretty irrelevant to its horizontal motion along a flat road, is now working against you when you try to get up the hill: you have to raise its weight many metres against the gravitational pull of the earth. This is why racing-bike manufacturers make a big deal of how light their bikes are. Obviously, the weight of an object can make a big difference to the ease of moving it; but our bicycle example illustrates that this difference depends on what kind of motion we're talking about.

Now, imagine that you wanted to discover whether the terrain between two towns, let's call them town A and town B, was flat or hilly, but were unable to travel between the towns yourself. A possible strategy emerges when you discover that a postal service exists between the towns manned by postmen who ride either a light or a heavy bicycle. To discover whether the intervening terrain is flat or hilly you need only post a set of identical packages between the towns, sending half via postmen riding the light bicycles and the rest by postmen on the heavier ones. If you discover that all the packages take about the same time to be delivered then you can conclude that the terrain between the two towns is probably quite flat; but if all the packages that arrived on heavy bicycles took much longer, you will conclude that the terrain between A and B is probably hilly. Your cycling postmen thereby act as probes of the unknown terrain.

Atoms of each chemical element come, like bicycles, in groups of different weights. Let's take hydrogen as our example since it is both the simplest atom and the one of most interest to us here. An element is determined by the number of protons it has in its nucleus (along with the corresponding equal number of electrons surrounding the nucleus). So, hydrogen has one proton in its nucleus, helium has two, lithium has three, and so on. But the nuclei of atoms also contain another type of particle: the neutron, which we met in chapter 1 when discussing the fusion of hydrogen nuclei inside the sun. Adding neutrons to the nucleus makes the atom heavier and therefore changes its physical attributes. Atoms of a particular element that have different numbers of neutrons are called *isotopes*. The normal isotope of hydrogen is the lightest one, consisting of just the single proton and electron. This is the most abundant form of hydrogen. But there are two rarer, heavier isotopes of hydrogen: deuterium (D), which has a neutron in addition to the proton in its nucleus, and tritium (T), which contains two neutrons.

Since the chemical properties of elements are determined mostly by the number of electrons their atoms contain, different isotopes of the same element, with different numbers of neutrons in their nuclei, will have very similar, but not identical, chemistries. The kinetic isotope effect involves measuring how sensitive a chemical reaction is to the changing of atoms from light to heavy isotopes, and is defined as the ratio of reaction rates observed with heavy and light isotopes. For example, if water is involved in a reaction then the hydrogen atoms in the H_2O molecules can be replaced with their heavier cousins, deuterium or tritium, to make D_2O or T_2O. Just like our cycling postmen, the reaction may or may not be sensitive to the changing weight of the atoms, depending on the route that the reactants take to be converted into products.

There are several mechanisms responsible for significant kinetic isotope effects, and one of them is quantum tunnelling, which, like cycling, is extremely sensitive to the mass of the particle that is trying to tunnel. Increasing the mass makes the particle's behaviour less wave-like and hence less likely to be able to seep through an energy barrier. So doubling the mass of the atom, for example changing from normal hydrogen to deuterium, causes its probability of quantum tunnelling to plummet.

Finding a big kinetic isotope effect may therefore be evidence that the reaction mechanism – the route between reactants and products – involves quantum tunnelling. However, it would not be conclusive since the effect might be attributable to some classical (non-quantum-driven) chemistry. But if quantum tunnelling is involved, then the reaction should also show a peculiar response to temperature: its rate should plateau out at low temperatures, just as DeVault and Chance had demonstrated for electron tunnelling. This is precisely what Klinman and her team discovered for the ADH enzyme; and the result provided strong evidence that quantum tunnelling was involved in the reaction mechanism.

Klinman's group has gone on to amass substantial evidence that proton tunnelling occurs commonly in many enzymatic reactions at the kinds of temperature at which life operates. Several other groups, such as that of Nigel Scrutton at the University of Manchester, have performed similar experiments with other enzymes and demonstrated kinetic isotope effects that point strongly towards quantum tunnelling.[10] Yet how enzymes maintain quantum coherence to promote tunnelling remains a very controversial topic. It has been known for some time that enzymes are not static, but are constantly vibrating during their reactions. For example, the *jaws* of the collagenase enzyme open and close every time they break a collagen bond. It was thought that these motions were either incidental to the reaction mechanism or were involved in capturing the substrates and bringing all the reactive atoms into the correct alignment. However, quantum biology researchers now claim that these vibrations are so-called 'driving motions' whose primary function is to bring atoms and molecules into close enough proximity to allow their particles (electrons and protons) to quantum tunnel.[11] We will be returning to this topic, one of the most exciting and fast-moving fields of quantum biology, in the last chapter of the book.

So does this establish the quantum in quantum biology?

Enzymes have made and unmade every single biomolecule inside every living cell that lives or has ever lived. Enzymes are as close as anything to the vital factors of life. So the discovery that some, and possibly all, enzymes work by promoting the dematerialization of particles from one point in space and their instantaneous materialization in another provides us with a novel insight into the mystery of life. And while there remain many unresolved issues related to enzymes that need to be better understood, such as the role of protein motions, there is no doubt that quantum tunnelling plays a role in the way they work.

Even so, we should address a criticism made by many scientists who accept the findings of Klinman, Scrutton and others, but nevertheless claim that quantum effects have as relevant a role in biology as they have in the workings of a steam train: they are always there but are largely irrelevant to understanding how either system works. Their argument is often positioned within a debate about whether or not enzymes *evolved* to take advantage of quantum phenomena such as tunnelling. The critics argue that the appearance of quantum phenomena in biological processes is inevitable given the atomic dimensions of most biochemical reactions. To a certain extent, they are right. Quantum tunnelling is not magic; it has been taking place in the universe since its birth. It is certainly not a trick that was somehow 'invented' by life. Yet we would argue that its appearance in enzyme activity is far from inevitable, given those hot, wet and busy conditions inside living cells.

Remember that living cells are extraordinarily crowded places, crammed with complex molecules in a state of constant agitation and turbulence, similar to that billiard-ball-like molecular motion we explored in the last chapter that is responsible for driving steam trains up hillsides. If you remember, it is this kind of random motion that scatters and disrupts the delicate quantum coherence and makes our everyday world appear 'normal' to us. Quantum coherence would not be expected to survive within this molecular turbulence, so the discovery that quantum effects, such as tunnelling, manage to persist in the sea of molecular agitation that is a living cell is very surprising. After all, it was only a decade or so ago that most scientists dismissed the idea that tunnelling and other delicate quantum phenomena could be taking place in biology. The fact that they have been found in these habitats suggests that life takes special measures to capture advantages provided by the quantum world *to make its cells work*. But what measures? How does life keep that enemy of quantum behaviour, decoherence, at bay? This is one of the biggest mysteries of quantum biology, but one that is slowly being unravelled, as we will discover in the final chapter.

But before we move on, we must return to where we left our nanosubmarine: at the active site of the collagenase enzyme inside the disappearing tail of a tadpole. We quickly exit the active site as the jaws reopen, allowing the broken collagen chain (and us!) to break free, and leave the clam-like enzyme ready to clip the next peptide bond in the chain. We then take a short cruise through the rest of the tadpole's body to witness the orderly activities of a few of the other enzymes that are equally vital for life. Following the migrating cells out of the shrinking tail and into the developing rear limbs, we witness new collagen fibres being laid down, like new railroad tracks, to support the building of the adult frog's body, often from cells migrating out of the disappearing tail. These new fibres are being built by enzymes that capture the amino-acid subunits released by collagenase and bolt them together again into new collagen fibres. Although we do not have the time to dive into these enzymes, within their active sites are the same kind of choreographed motions that we witnessed inside the collagenase enzyme, but now performing the reverse reaction. Elsewhere, all the biomolecules of life – fats, DNA, amino acids, proteins, sugars – are being made and unmade by different enzymes. Also, every action the growing frog performs is similarly mediated by enzymes. For example, when the animal spots a fly, the nerve signals that carry the message from its eyes to its brain are mediated by a group of neurotransmitter enzymes crammed into nerve cells. As it flings out its tongue, the muscular contractions that draw in the fly are driven by another enzyme, called myosin, crammed into muscle cells, which cause those cells to contract. When the fly enters the frog's stomach, a whole battery of enzymes is released to speed its digestion and release its nutrients so they can be absorbed. Yet more enzymes transform those nutrients into frog tissue, or capture their energy via respiratory enzymes within its mitochondrial cell organelles.

Every *vital* activity of frogs and other living organisms, every process that keeps them – and us – alive, is accelerated by enzymes, the engines of life, whose extraordinary catalytic power is provided by their ability to choreograph the motions of fundamental particles and thereby dip into the quantum world to harness its strange laws.

But tunnelling isn't the only potential advantage provided to life by quantum mechanics. In the next chapter we will discover that the most important chemical reaction in the biosphere involves another trick of the quantum world.

4

The quantum beat

The substance of a tree is carbon and where did that come from?
That comes from the air; it's carbon dioxide from the air. People
look at trees and they think it [the substance of the tree] comes out
of the ground; plants grow out of the ground. But if you ask 'where
does the substance come from' you find out . . . the trees come out of
the air . . . the carbon dioxide and the air goes into the tree and it
changes it, kicking out the oxygen . . . We know that the oxygen and
carbon [in carbon dioxide] stick together very tight . . . how does the
tree manage to undo that so easily? . . . It is the sunlight that comes
down and knocks this oxygen away from the carbon . . . leaving the
carbon, and water, to make the substance of the tree!

Richard Feynman[1]

THE MASSACHUSETTS Institute of Technology, better known as
MIT, is one of the world's scientific powerhouses. Founded in
1861 in Cambridge, Massachusetts, it boasts nine current Nobel
laureates among its one thousand professors (as of 2014). Its alumni
include astronauts (one-third of NASA's space flights were manned
by MIT graduates), politicians (including Kofi Annan, former
Secretary-General of the United Nations and winner of the 2001
Nobel Peace Prize), entrepreneurs such as William Reddington
Hewlett, co-founder of Hewlett-Packard – and, of course, lots of
scientists, including the Nobel Prize-winning architect of quantum

electrodynamics, Richard Feynman. Yet one of its most illustrious inhabitants is not human; it is in fact a plant, an apple tree. Growing in the President's Garden in the shadow of the institute's iconic Pantheonesque dome is a cutting from another tree kept at England's Royal Botanic Gardens, which is a direct descendant of the actual tree under which Sir Isaac Newton supposedly sat when he observed the falling of his famous apple.

The simple yet profound question that Newton had been contemplating sitting under a tree at his mother's Lincolnshire farm three and a half centuries ago was: *why do apples fall?* It may seem churlish to suggest that his answer, one that revolutionized physics and indeed all of science, could be inadequate in any way; but there is an aspect of that famous scene that went unnoticed by Newton and has gone unremarked upon ever since: what was the apple doing up in the tree in the first place? If the apple's accelerated descent to the ground was puzzling, then how much more inexplicable was the bolting together of Lincolnshire air and water to form a spherical object perched in the branches of a tree? Why did Newton wonder about the comparatively trivial matter of the pull of the earth's gravity on the apple and overlook entirely the utterly incomprehensible puzzle of the fruit's formation in the first place?

One factor that might explain Isaac Newton's lack of curiosity about this was the predominant seventeenth-century view that although the brute mechanics of all objects, including living ones, might be accounted for by physical laws, their peculiar inner dynamic (dictating, among other things, how apples grow) was driven by that vital force or *élan vital* which flowed from a supernatural source beyond the reach of any godless mathematical equation. But, as we have already discovered, vitalism was blown away by subsequent advances in biology, genetics, biochemistry and molecular biology. No serious scientist today doubts that life can be accounted for within the sphere of science; but there remains a question mark over which of the sciences can best provide that account. Despite the alternative claims of scientists such as Schrödinger, most biologists still believe that the classical laws are sufficient, with Newtonian forces acting upon ball and stick biomolecules that behave like, well, balls and sticks. Even Richard Feynman, one of Schrödinger's intellectual successors, described photosynthesis (in the passage quoted at the head of this chapter) in strictly classical terms with 'sunlight that comes down and knocks this oxygen away from the carbon', with light acting like some kind of golf club able to whack the oxygen golf ball out of the carbon dioxide molecule.

Molecular biology and quantum mechanics developed in parallel, rather than cooperatively. Biologists hardly attended physics lectures and physicists paid little attention to biology. But in April 2007, a group of MIT-based physicists and mathematicians who worked in a rather esoteric area called quantum information theory were enjoying one of their regular journal clubs (with each member taking a turn at presenting a new paper they had found in the scientific literature) when one of the group arrived with a copy of the *New York Times* carrying an article which suggested plants were *quantum computers* (more on these remarkable machines in chapter 8). The group exploded into laughter. One of the team, Seth Lloyd, recalled first hearing about this 'quantum hanky-panky'. 'We thought that was really hysterical . . . It's like, "Oh my God, that's the most crackpot thing I've heard in my life!"'[2] The cause of their incredulity was the fact that many of the brightest and best-funded research groups in the world had spent decades trying to figure out how to build a quantum computer, a machine that could carry out certain calculations much faster and far more efficiently than the most powerful computers available in the world today (since, rather than relying on digital bits of information that are either 0 or 1, it would allow them to be both 0 and 1 simultaneously and therefore be able to pursue all possible calculations at once – the ultimate in parallel processing). The *New York Times* article was claiming a humble blade of grass was able to perform the kind of quantum trickery that lay at the heart of quantum computing. No wonder these MIT researchers were incredulous. They might not be able to build a working quantum computer but, if the article was right, they could eat one in their lunchtime salad!

Meanwhile, not far from the room where the MIT journal club were laughing their quantum socks off, a photon of light travelling at 186,000 miles per second was hurtling towards a tree with a famous pedigree.

The central mystery of quantum mechanics

We'll return to that photon and tree shortly, and how they might be related to the quantum world, but first you need to be introduced to a beautifully simple experiment that highlights just how weird the quantum world really is. While we will go to great lengths to explain as best as we can just what is meant by notions such as 'quantum superposition', nothing really hits the message home better than the famous two-slit experiment, which we will describe here.

What the two-slit experiment delivers is the simplest and starkest demonstration that, down in the quantum world, *everything is different*. Particles can behave like waves spread out across space and waves can sometimes act like individual localized particles. You've already encountered this wave–particle duality: in the opening chapter, as the peculiarity that is necessary to account for how the sun generates its energy; and in chapter 3, where we saw how the wave properties of electrons and protons allow them to seep though energy barriers inside enzymes. In this chapter you will discover that wave–particle duality is also involved in the most important biochemical reaction in the biosphere: the conversion of air, water and light into plants, microbes and, indirectly, all the rest of us. But first we must discover how the outlandish idea that particles can be in many places at once is forced upon us by one of the simplest, most elegant, but most far-reaching experiments ever performed: the experiment that, according to Richard Feynman, 'has in it the heart of quantum mechanics'.

Be warned, however, that what will be described here will seem impossible, and you may feel certain that there just has to be a more rational way of explaining what is going on. You may be left wondering where the sleight of hand is in what seems to be a magic trick. Or you may assume that the experiment is mere theoretical speculation dreamt up by scientists lacking the imagination to comprehend the workings of nature. But neither of these explanations is correct. The two-slit experiment doesn't make (common) sense, but it is real and has been performed thousands of times.

We will describe the experiment in three stages; the first two will merely set the scene, so that you can then appreciate the baffling results of the third, and main, stage.

First, a beam of monochromatic light (consisting of a single colour, or wavelength) is shone on a screen with two narrow slits that allow some of the light to pass through both slits onto a second screen (figure 4.1). By carefully controlling the width of the slits, their distance apart from each other, and the distance between the two screens, we can create a sequence of light and dark bands on the second screen, known as an interference pattern.

Interference patterns are the signature of waves and are easy to see in any wavy medium. Toss a pebble into a still pond and watch as a set of concentric circular waves travel outwards from the splash point. Toss two pebbles into the same pond and each will generate its own expanding concentric waves, but where the waves from the two pebbles overlap you will see an interference pattern (figure 4.2). Wherever the peak of one wave meets the trough of another they cancel out, resulting in no wave at those points. This is called destructive interference. Conversely, where two peaks or two troughs meet, they reinforce each other, generating twice the wave: this is called constructive interference. This pattern of wave cancellation and reinforcement can be produced in any wavy medium. In fact, it was the English physicist Thomas Young's demonstration of interference of light beams in an early version of the two-slit experiment performed over two centuries ago that convinced him and most other scientists that light was indeed a wave.

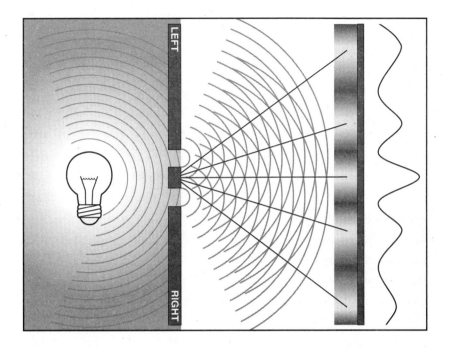

Figure 4.1: The two-slit experiment, stage 1. When monochromatic light (having a specific wavelength) is shone onto the two slits, each slit then acts as a new source of light on the other side and, because of its wave-like nature, the light spreads out (diffracts) as it squeezes through each slit so that the circular waves overlap and interfere with each other, leading to light and dark fringes on the back screen.

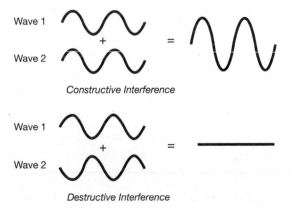

Figure 4.2: Constructive and destructive interference of waves.

The interference shown in the two-slit experiment is due first to the way light waves pass through both slits and then spread out, a property of waves known as diffraction, so that the beams emerging from the slits overlap and merge, just as water waves do, before hitting the back screen. At certain points on the screen the light waves emanating from the two slits will arrive *in phase*, with peaks and troughs marching in step, either because they have covered the same distance to the screen or because the difference in the distance they travelled is equal to a multiple of the distance between their peaks. Where this happens, the crests and troughs of the waves combine to form higher crests and lower troughs: constructive interference. The fused waves create high-intensity light at these points and hence a bright band on the screen. But at other points, the light from the two slits arrives *out of phase*, at the point where the crest of one wave meets the trough of another. At these points the waves cancel out, resulting in a dark band on the screen: destructive interference. In between these two extremes the combination is neither completely 'in phase' nor completely 'out of phase' and some light survives. We therefore don't see a sharp sequence of light and dark bands on the screen but a smooth variation in intensity, between what are known as maxima and minima in the interference pattern. This appropriately wave-like smooth variation in intensity is a key indicator of wave phenomena. One familiar example of this can be found with sound waves: a musician tuning an instrument listens for the 'beats'* that occur when one note is very close in frequency to another so that as they travel to the musician's ear they sometimes arrive in phase and sometimes out of phase. This variation in their combined pattern generates an overall sound that periodically rises and falls in volume. This smooth variation in the intensity of the sound is due to interference between two separate waves. Note that these beats are an entirely classical example that requires no quantum explanation.

---

* These are fluctuations in volume – a kind of pulse – created by two notes that are almost the same frequency and thereby nearly in tune. This use of the term 'beat' should not be confused with the more common use of 'beat' in music to mean its rhythm.

A key factor in the two-slit experiment is that the beam of light hitting the first screen must be monochromatic (consisting of a single unique wavelength). In contrast, white light, such as that emitted by a normal light bulb, is composed of many different wavelengths (all the colours of the rainbow), so that the waves will arrive at the screen in a higgledy-piggledy fashion. In this case, although peaks and troughs will still interfere with each other, the resulting pattern will be so complex and so smeared out that no distinct bands will be seen. In a similar way, although it is easy to generate an interference pattern when we drop two pebbles into a pond, a huge waterfall crashing into the pond generates so many waves that it is impossible to find any coherent interference pattern.

Now for stage two of the two-slit experiment, which we perform, not with light, but by firing bullets at the screen. The point here is that we are using solid particles rather than spread-out waves. Each bullet must of course pass through one or the other slit, not both. With enough bullets getting through we see that the back screen will have accumulated two bands of bullet holes corresponding to the two slits (figure 4.3). Clearly, we are not dealing with waves. Each bullet is an independent particle and has no relationship with any other bullet, so there is no interference.

Now for stage three: the quantum 'magic trick'. The experiment is repeated using atoms instead of bullets. A source that can produce a beam of atoms fires them at a screen with two appropriately narrow slits.* To detect the arrival of the atoms, the second screen has a photoluminescent coating that shows up as a tiny bright spot wherever a single atom hits it.

If common sense prevailed at the microscopic level, then atoms should behave like incredibly tiny bullets. We run the experiment first with just the left slit open and see a band of light spots on the back screen behind the open slit. There is a certain amount of

---

* The slits do indeed need to be very narrow and very close together. In the experiments carried out in the 1990s the screen was a sheet of gold foil and the slits were of the order of a single micrometre (a thousandth of a millimetre) wide.

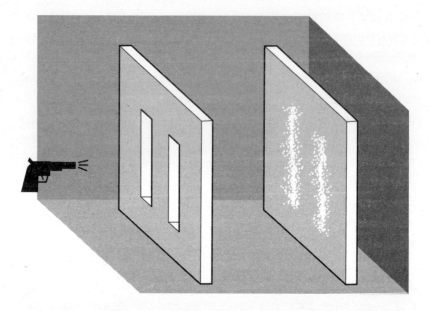

Figure 4.3: The two-slit experiment, stage 2. Unlike the wave behaviour we saw with light, firing a stream of bullets at the slits shows particle-like behaviour. Each bullet that gets through to the back screen must have gone through one or other slit, but not both (assuming, of course, that the middle screen is thick enough to block any bullets that miss the slits). Rather than multiple banded interference, the pattern on the back screen now shows an accumulation of bullets along just two narrow strips adjacent to each of the slits.

spreading of the spots that one might presume to result from some of the atoms bumping off the edges and being deflected rather than going cleanly through the slit. Next, we also open the right slit and wait for the spots to build up on the screen behind it.

If you were asked to predict the distribution of the bright spots and you knew nothing of quantum mechanics, then you would naturally guess that it would look very much like the pattern produced by the bullets; namely, that a band of spots would build up behind each slit, giving two distinct patches of light that are brightest in their centre and gradually fade away as we move out – as the atom 'hits' become rarer. You would also expect that the mid-point between the two bright patches would be dark, since it corresponds to a region of the screen that is hardest to reach for the atoms, whichever slit they manage to get through.

But this is not what we find. Instead, we see a very clear interference pattern of light and dark fringes, just like we did with light. The brightest part of the screen, believe it or not, is in the centre of the screen: the very patch we would not expect many atoms to be able to reach (figure 4.4). In fact, with the right distance between the slits and the right distance between the two screens, we can make sure that the bright region on the back screen (the area atoms were able to reach with just one slit open) will now be dark (no atoms arriving there) when we open up the second slit. How can opening up another

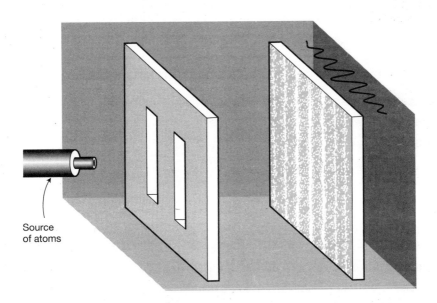

Source
of atoms

Figure 4.4: The two-slit experiment, stage 3. Replacing the bullets with atoms from a source that can fire them at the slits (of course, in each stage the width and separation of the slits are chosen appropriately) we see the wave-like interference pattern appearing again. Despite each atom hitting the back screen at a localized point, indicating its particle nature, they cluster together in bands, just as was seen with light. So what is it that is going through both slits at once, and without which we would not see the multiple interference fringes?

slit, which should only allow more atoms through, prevent atoms reaching certain regions of the screen?

Let us see if we can explain what is going on using simple common sense and avoid appealing to quantum mechanics just yet. Suppose that, despite each atom being a tiny localized particle – after all, every atom hits the screen at a single point – the sheer number of atoms involved, all colliding and interacting with each other in a particular coordinated way, produces a pattern with the *appearance* of interference. After all, we know that water waves are in fact composed of lots of molecules of water that, on their own, would not be expected to be wavy. It is the coordinated motion of trillions of water molecules that produces the wave-like properties, not each molecule individually. Perhaps the atom gun extrudes a coordinated flow of atoms, rather like a wave machine in a swimming pool.

To test the coordinated atom theory, we repeat the experiment, but now send the atoms through *one at a time*. We fire the atom gun and wait for the appearance of a spot of light on the back screen before firing it again, and so on. Initially, common sense seems to prevail. Each atom that manages to get through the slits leaves just one tiny localized spot of light somewhere on the screen. It seems that atoms leave the gun as bullet-like particles and arrive at the screen as particles. Surely, in between gun and screen they must similarly behave as particles? But now the quantum rabbit comes out of the hat. As the spots, each one recording the arrival of a single bullet-like atom, gradually build up on the screen, the light and dark interference pattern gradually emerges once again. With the atoms now travelling through the instrument one at a time, we can no longer argue that there is any collective behaviour of lots of atoms bumping into and interacting with each other. These aren't like water waves. And here again, we must confront the counterintuitive result that there are places on the back screen where atoms could land when only one of the slits was open, and yet are completely dark if the second slit is also open, despite its opening providing an additional route for the atom to reach the screen. It seems as though an atom passing through one slit must somehow be *aware* whether or not the other slit is open, and act accordingly!

To recap, each atom leaves the gun as a tiny localized particle and arrives at the second screen also as a particle, as is evident from the tiny flash of light when it arrives. But in between, as it encounters the two slits, there is something mysterious going on akin to the behaviour of a spread-out wave that gets split into two components, each emerging from a slit and interfering with the other on the far side. How else can a single atom be *aware* of the state (whether open or closed) of both slits at the same time?

Suspecting sleight of hand somewhere, let's see if we can catch the atoms out by lying in wait behind the slits. This can be achieved by setting up a detector, behind the left slit, say, so that it registers a 'signal' (maybe a beep) whenever an atom passes through that slit on its way to the screen.* We can also place a second detector over the right slit to catch atoms that pass through that slit. Now, if an atom passes through one slit or the other, we will hear a beep from either the left or the right detector; but if the atom manages to somehow unwrap its bullet-like nature and go through both slits, then both detectors will beep at the same time.

What we now find is that, with each firing of the atom gun that is accompanied by the appearance of a bright dot on the screen, either the left or the right detector beeps, never both. Surely we now have proof at last that the interfering atoms do indeed go through either one slit or the other, but not both simultaneously. But, be patient and keep watching the screen. As lots of individual flashes of light build up and coalesce, we see that what is produced is no longer an interference pattern. In its place are just two bright bands, indicating the collection of a pile of atoms behind each slit, just like we had in the experiment with bullets. The atoms are now behaving like

---

*We assume here that the detector has 100% efficiency and will definitely fire if an atom passes through the slit it is watching, and yet does not interfere with the path of the atom. Of course, in practice this is not possible since we unavoidably disturb the passage of the atom through the act of observation, as we are about to see.

conventional particles throughout the experiment. It is as though each atom behaves like a wave when it is confronted by the slits, *unless* it is being spied upon, in which case it innocently remains as a tiny particle.

Maybe the presence of the detector is causing a problem, perhaps upsetting the strange and delicate behaviour of the atoms going through the slits. Let's test this by removing one of the detectors, say the detector over the right slit. We can still get the same information from this arrangement because when we fire the gun, hear a beep and see a bright spot on the screen, we will know the atom must have gone through the left slit; when we fire the gun, don't hear a beep but do see a bright dot, then we'll know that the atoms must have reached the screen via the right slit. We can now know whether the atoms have gone through the left or right slit, but we are only 'disturbing' one of the routes. If the detector itself was causing the problem then we would expect those atoms that triggered the beep to behave like bullets but those that didn't (and went through the right slit) to behave like waves. Maybe now we'll see a mixture of a bullet-like pattern (from atoms going through the left slit) and an interference pattern (from atoms going through the right slit) on the screen.

But we don't. With this arrangement we still don't see any interference pattern. Only the bullet-like pattern of dots is seen behind each slit on the screen. It seems that the mere presence of a detector that can register the location of an atom is enough to destroy its wave-like behaviour, even if that detector is some distance away from the atom's path through the other slit!

Perhaps the physical presence of the detector over the left slit is sufficient to influence the path of atoms passing through it, rather like a large boulder changing the flow of water in a fast-moving stream. We can test this by switching off the left detector. It's still there, so we would expect its influence to be pretty much the same. But now, with the detector present but switched off, the interference pattern builds up on the screen once again! All the atoms going through the experiment have gone back to behaving as waves. How is it that atoms behaved as particles when the detector over the left slit was switched on, but as soon as it was switched off they behaved like waves? How does a particle going through the right slit *know* that the detector over the left slit is switched on or off?

It is at this stage that you have to leave common sense behind. Now we have to confront the wave–particle duality of tiny objects such as atoms, electrons or photons, that behave like a wave when we do not have information about which slit they went through, but like a particle when we observe them. This is the process of observation or measurement of quantum objects that we first met in chapter 1 when considering Alain Aspect's demonstration of quantum entanglement in separated photons. You will remember that Aspect's team measured their photons by passing them through a polarized lens that destroyed their entangled state – which is an aspect of their wave nature – forcing them to *choose* a single classical polarization direction. In a similar way, the measurement of atoms passing through the two-slit experiment forces them to choose whether to go through the left or the right slit.

Quantum mechanics does in fact provide us with a perfectly logical explanation of this phenomenon; but it is only an explanation of what we observe – the result of an experiment – not of what is going on when we are not looking. But since all we have to go on is what we can see and measure, maybe it makes no sense to ask for more. How can we assess the legitimacy or truth of an account of a phenomenon that we can never, even in principle, check? As soon as we try, we alter the outcome.

The quantum interpretation of the two-slit experiment is that at any given moment in time, each atom must be described by a set of numbers that define its probabilistic location in space. This is the quantity we introduced in chapter 2 as the *wave function*. There we described it as being similar to the idea of tracking a crime wave spreading through a city by assigning probabilities to burglaries taking place in different districts. In a similar way, the wave function describing an atom going through the two slits tracks the likelihood of finding it anywhere in the apparatus at any given time. But, as we emphasized earlier, whereas a burglar must have a single location in space and time, and the 'crime probability' wave describes only our lack of knowledge of where he actually is, in contrast, the wave function of the atom in the two-slit experiment is *real* in the sense that it represents the physical state of the atom itself, which really doesn't have a specific location unless we measure it and is, until then, everywhere at once – with varying probability, of course, so that we are unlikely to find the atom in places where its wave function is small.

So instead of individual atoms going through the two-slit experiment we have to consider the wave function travelling from source to back screen. On encountering the slits, the wave function splits in two, with each half going through one of the slits. Note that what we are describing here is the way an abstract *mathematical* quantity changes in time. It is pointless to ask what is *really* going on, since we would have to look to check. But as soon as we try to do so we alter the outcome. Asking what is *really* going on between observations is like asking whether your fridge light is on before you open the fridge door: you can never know because as soon as you peek you change the system.

The question then arises: when does the wave function 'become' a localized atom once again? The answer is: when we try to detect its location. When such a measurement takes place, the quantum wave function *collapses* to a single possibility. Once again, this is very different from the burglar situation where the uncertainty about his whereabouts suddenly collapsed to a single point after he was nabbed by the police. In that case, it was only our information about the burglar's whereabouts that was affected by the detection. He was always only ever in one place at any given time. Not so for the atom; in the absence of any measurement, the atom really is everywhere.

So, the quantum wave function calculates the probability of detecting the atom at a specific location, *were we to carry out a measurement of its position at that time*. Where the wave function is large before measurement, the resulting probability of finding the atom there will be high. But where it is small, perhaps due to destructive wave interference, the probability of finding the atom there when we decide to look will be correspondingly small.

We can imagine following the wave function describing the single atom as it leaves the source. It behaves just like a wave that flows towards the slits, so, at the level of the first screen, it will be of equal amplitude in each slit. If we place a detector on one of the slits, then we should expect equal probabilities: 50 per cent of the time we will detect the atom at the left slit and 50 per cent of time we will detect it at the right slit. But – and this is the important bit – if we don't try to detect the atom at the level of the first screen then the wave function flows through both slits without collapsing. Thereafter, in quantum terms we can talk of a wave function describing a single atom that is in a superposition: of its being in two places at the same time, corresponding to its wave function going through both the left and right slits simultaneously.

On the other side of the slits, each separated piece of the wave function, one from the left and one from the right slit, spreads out again and both form sets of mathematical ripples that overlap, at some points reinforcing and at other points cancelling each other's amplitude. The combined effect is that the wave function now has the pattern characteristic of other wave phenomena, such as light. But bear in mind that this now complicated wave function is still describing only a single atom.

At the second screen, where a measurement of the position of the atom finally takes place, the wave function allows us to calculate the probability of detecting the particle at different points along the screen. The bright patches on the screen correspond to those positions where the two parts of the wave function, coming from the two slits, reinforce each other, and the dark patches correspond to those positions where they cancel each other out to generate a zero probability for atoms being detected at these positions.

It is important to remember that this reinforcement and cancellation process – quantum interference – takes place even when only a single particle is involved. Remember that there are regions of the screen that atoms, fired one at a time, could reach with just one slit open but that were no longer reachable when both slits are open. This only makes sense if each atom released from the atom gun is described by a wave function that can explore both paths simultaneously. The combined wave function with its regions of constructive and destructive interference cancels out the probability of the atom being found in some positions on the screen that it would reach if only one slit were open.

All quantum entities, whether fundamental particles or the atoms and molecules composed of these particles, display coherent wave-like behaviour so that they can interfere with themselves. In this quantum state they can exhibit all the weird quantum behaviours, such as being in two places at once, spinning in two directions at once, tunnelling through impenetrable barriers or possessing spooky entangled connections with a distant partner.

But then, why can't you or I, ultimately composed as we are of quantum particles, be in two places at once; something that would certainly be extremely useful on a busy day? The answer on one level is very simple: the bigger and more massive a body is, the smaller will its wave-like nature be, and something the size and mass of a human, or indeed anything large enough to be visible with the naked eye, will have a quantum wavelength so tiny as to have no measurable effect. But more deeply, you can think of each atom in your body as being observed, or measured, by all the other atoms around it, so that any delicate quantum properties it might have are very quickly destroyed.

What, then, do we actually mean by 'measurement'? We have already briefly explored this question in chapter 1, but we must now take a closer look since it is central to the question of how much 'quantum' there is in quantum biology.

Quantum measurement

For all its success, quantum mechanics tells us nothing about how to take the step from the equations that describe how an electron, say, moves around an atom to what we see when we make a specific measurement of that electron. For this reason, the founding fathers of quantum mechanics came up with a set of ad hoc rules that became an addendum to the mathematical formalism. They are known as the 'quantum postulates' and provide a sort of instruction manual on how to translate the mathematical predictions of the equations into tangible properties we can observe, such as the position or energy of an atom at any given moment.

As for the actual process itself whereby an atom instantaneously stops being 'over here *and* over there' and is just 'over here' when we look, no one really knows what goes on and most physicists have been happy to adopt the pragmatic view that it 'just happens'. The problem is that this requires an arbitrary distinction to be made between the quantum world, where weird stuff happens, and our everyday macroworld where objects behave 'sensibly'. A measuring device that detects an electron has to be part of this macroworld. But *how* and *why* and *when* this measurement process takes place was never clarified by the founders of quantum mechanics.

During the 1980s and 1990s, physicists came to appreciate what must be happening when an isolated quantum system, such as a single atom in the two-slit experiment, with its wave function existing in its superposition of being in two places at once, interacts with a macroscopic measuring device, say one placed on the left slit. It turns out that detecting the atom (and note here that even *not* detecting the atom is regarded as a measurement, as that means it must have gone through the other slit) causes the atom's wave function to interact with all the trillions of atoms in the measuring device. This complex interaction causes the delicate quantum coherence to leak away very quickly and be lost in the incoherent noise of its surroundings. This is the process called decoherence that we have already met in chapter 2.

But decoherence does not need a measuring device to come into effect. It is taking place all the time inside every single classical object as its quantum constituents – the atoms and molecules – undergo thermal vibrations and get buffeted around by all the surrounding atoms and molecules, so that their wave-like coherence is lost. In this way we can think of decoherence as the means by which all the material surrounding any given atom, say – what is referred to as its environment – is constantly *measuring* that atom and forcing it to behave like a classical particle. In fact, decoherence is one of the fastest and most efficient processes in the whole of physics. And it is because of this remarkable efficiency that decoherence evaded discovery for so long. It is only now that physicists are learning how to control and study it.

Returning to our analogy of throwing pebbles into water, when we threw them into a still pond it was easy to see their overlapping waves interfering with each other. But try throwing those same pebbles into the base of Niagara Falls. The hugely complex and chaotic nature of the water now immediately wipes out any interference pattern generated by the pebbles. This turbulent water is the classical equivalent of the random molecular motion surrounding a quantum system, resulting in instant decoherence. Most environments are, at a molecular level, just as turbulent as the waters at the base of Niagara Falls. Particles within materials are constantly being jostled and bumped around by their environment (other atoms, molecules or photons of light).

At this point we should clarify some of the terminology we are using in this book. We talk about atoms being in two places at once, behaving like spread-out waves and existing in a superposition of two or more different states at once. By way of making things easier for you, the reader, we can settle on a single term that encompasses all these concepts: that of quantum 'coherence'. Thus, when we refer to 'coherent' effects we mean something is behaving in a quantum mechanical way, exhibiting wave-like behaviour or doing more than one thing at the same time. Thus, 'decoherence' is the physical process whereby coherence is lost and the quantum becomes classical.

Quantum coherence is normally expected to be very short-lived unless the quantum system can be isolated from its surroundings (fewer jostling particles) and/or cooled to a very low temperature (much less jostling) to preserve the delicate coherence. In fact, to demonstrate interference patterns with single atoms, scientists pump all the air out of the apparatus and cool their equipment down to very close to absolute zero. Only by taking these extreme steps can they maintain their atoms in a quiet quantum coherent state for long enough to demonstrate the interference patterns.

The issue of the fragility of quantum coherence (keeping the wave function from collapsing) is of course the principal challenge to the MIT group whom we met in the opening paragraphs of this chapter, and their colleagues around the world, in their quest to build a quantum computer; and this was why they were so sceptical about the *New York Times* claim that plants were quantum computers. Physicists come up with all sorts of clever and expensive stratagems to shield the quantum world inside their computers from the coherence-destroying outside environment. So the idea that quantum coherence could be maintained in the hot, wet and molecularly turbulent environment inside a blade of grass was understandably thought to be crazy.

However, we now know that down at the molecular level, many important biological processes can indeed be very fast (of the order of trillionths of a second) and can also be confined to short atomic distances – just the sort of length and timescales where quantum processes like tunnelling can have an effect. Thus, although decoherence can never be entirely prevented, it may be kept at bay for just long enough to be biologically useful.

Voyage to the centre of photosynthesis

Glance up at the sky for one second and a column of light 186,000 miles long descends into your eye. In that same second, the earth's plants and photosynthetic microbes harvest the solar light column to make about 16,000 tonnes of new organic matter in the form of trees, grass, seaweed, dandelions, giant redwoods and apples. Our aim in this section is to discover how this first step in the transformation of inanimate matter into nearly all of the biomass on our planet actually works; and our exemplar transformation will be the conversion of New England air into an apple on Newton's tree.

To see this process in action, we will borrow once again the nano-technology submarine that we used to explore enzyme action in the last chapter. Once you've climbed aboard and flicked the miniaturization switch, you launch the craft skyward, up into the foliage of the tree, where you alight upon one of its expanding leaves. The leaf continues to expand until its furthest edges are lost beyond the horizon and its apparently smooth surface becomes an irregular platform paved with rectangular green bricks pock-marked by paler round blocks, each penetrated by a central pore. The green bricks are called epidermis cells and the round blocks are called stomata: their job is to allow air and water (the substrates of photosynthesis) to pass through the surface of the leaf into its interior. You guide the craft over to the nearest stoma and, when the vessel is only a micron (a millionth of a metre) in length, you lower its prow to dive through the pore and emerge within the green and bright interior of the leaf.

Once inside, you come to rest within the roomy and rather still space of the leaf's interior, floored by rows of boulder-like green cells and roofed by thick cylindrical cables. The cables are the *veins* of the leaf, which either carry water from the roots to the leaf (xylem vessels) or transport newly made sugars from the leaf to the rest of the plant (phloem vessels). As you shrink further, the face of the boulder-like cell expands in all directions until it appears to be the size of a football pitch. At this scale – you are now about 10 nanometres tall, or one hundred-thousandth of a millimetre – you can see that its surface is turfed with a ropy mesh of cords, rather like a thick jute rug. This corded material is the *cell wall*, which is a kind of cellular exoskeleton. Your nanosubmarine is armed with instruments that you use to hack a path down through this ropy rug, revealing a waxy underlay, the cell membrane, which is the final water-impermeable barrier between the cell and its external environment. A closer inspection reveals that it is not entirely smooth, but pockmarked by water-filled holes. These membrane channels are called *porins* and they are the cell's plumbing, allowing nutrients in and waste products out. To enter the cell you need only wait alongside one of the porins until it has expanded sufficiently for you to dive into the cell's watery interior.

Once through the porin channel you can immediately see that the inside of a cell is very different from its exterior. Instead of majestic columns and wide-open spaces, this interior is crowded and somewhat messy. It also looks like a very busy place! The watery fluid filling the cell, known as the *cytoplasm*, is thick and viscous; in places it's more like a gel than a liquid. And suspended in the gel are thousands of irregular globular objects that appear to be in a state of constant internal motion. These are protein enzymes, like those we met in the last chapter, responsible for conducting the cells' metabolic processes, breaking down nutrients and making biomolecules such as carbohydrates, DNA, protein and fats. Many of these enzymes are tethered to a network of cables (the cell's *cytoskeleton*) which, rather like chair-lift cables, appear to be pulling numerous cargoes to various destinations within the cell. This transport network appears to emanate from several hubs, where the cables are anchored onto large green capsules. These capsules are the cell's *chloroplasts*, within which the central action of photosynthesis takes place.

You propel the submarine through the viscous cytoplasm. Progress is slow, but you eventually arrive at the nearest chloroplast. It lies beneath you like a huge green balloon. You can see that it, like the enclosing cell, is bounded by a transparent membrane through which great stacks of green coin-like objects are visible. These are the *thylakoids* and they are packed full of molecules of chlorophyll, the pigment that makes plants green. Thylakoids are the engines of photosynthesis that, when fuelled by photons of light, can bolt carbon atoms (absorbed from the carbon dioxide in the air) together to make the sugars that will go into our apple. To get a better view of this first step in photosynthesis, you steer the craft through one of the pores in the chloroplast membrane towards the topmost green coin of the thylakoid stack. Having reached your destination, you switch off the craft's engine, allowing the vessel to hover above this powerhouse of photosynthetic action.

Below you lies just one of the trillions of photosynthetic machines that manufacture the world's biomass. From your vantage point you can see that, as we discovered when examining enzyme machinery in the last chapter, although there are plenty of the billiard-ball-like turbulent molecular collisions going on all around you, there is also an impressive degree of order. The membranous surface of the thylakoid is studded with craggy green islands forested with tree-like structures terminating in antennae-like pentagonal plates. These

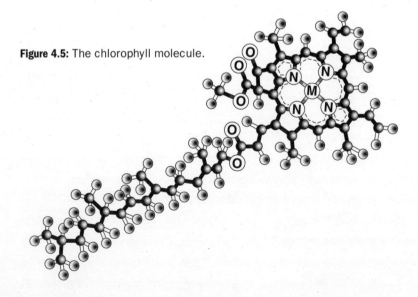

Figure 4.5: The chlorophyll molecule.

antennae plates are light-harvesting molecules called *chromophores*, of which chlorophyll is the most famous example, and it is these that perform the first crucial step of photosynthesis: capturing light.

Probably the second most important molecule on our planet (after DNA), chlorophyll is worth a closer look (figure 4.5). It is a two-dimensional structure made up of pentagonal arrays of mostly carbon (grey spheres) and nitrogen (N) atoms enclosing a central magnesium atom (M), with a long tail of carbon, oxygen (O) and hydrogen (white) atoms. The magnesium atom's outermost electron is only loosely bound to the rest of the atom and can be knocked into the surrounding carbon cage by absorption of a photon of solar energy to leave a gap in what is now a positively charged atom. This gap, or electron hole, can be thought of in a rather abstract way as a 'thing' in itself: a positively charged hole. The idea is that we regard the rest of the magnesium atom as remaining neutral while we have created, through the absorption of the photon, a system consisting of the escaped negative electron and the positive hole it has left behind. This binary system is called an *exciton* (see figure 4.6) and can be

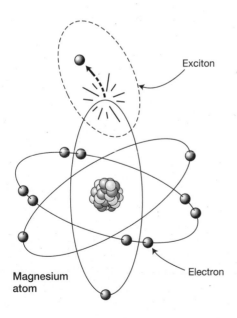

Figure 4.6: An exciton consists of an electron that has been knocked out of its orbit in an atom, together with the hole it leaves behind.

Exciton

Magnesium atom

Electron

thought of as a tiny battery with positive and negative poles capable of storing energy for later use.

Excitons are unstable. The electron and its hole feel an attractive electrostatic force pulling them together. If they recombine, the solar energy of the original photon is lost as waste heat. So, if the plant is to harness its captured solar energy, it has to transport the exciton very rapidly to a molecular manufacturing unit known as the *reaction centre*, where a process called charge separation takes place. Essentially, this involves stripping an energetic electron completely from its atom and transferring it to a neighbouring molecule, rather like the enzymatic action we observed in the last chapter. This process creates a more stable chemical battery (called NADPH) than an exciton that is used to drive the all-important photosynthetic chemical reactions.

But reaction centres are usually quite distant, in molecular terms (nanometre distances) from the excited chlorophyll molecules, so the energy has to be transferred from one antenna molecule to another within the chlorophyll forest to reach the reaction centre. This can happen thanks to the tightly packed nature of the chlorophyll. Molecules neighbouring the one that has absorbed the photon can themselves become excited, effectively inheriting the energy of the initially excited electron, which is then transferred to their own magnesium atom's electron.

The problem, of course, is which route this energy transfer should take. If it heads in the wrong direction, randomly hopping from one molecule to the next in the chlorophyll forest, it will eventually lose its energy rather than delivering it to the reaction centre. Which way should it turn? It doesn't have very long to find its way to its destination before the exciton expires.

Until recently, it was thought that this energy-hopping from one chlorophyll molecule to another was haphazard, essentially adopting the search strategy of last resort, known as a random walk. This is sometimes referred to as a 'drunken walk' because it resembles the path taken by an intoxicated drinker exiting a bar, wandering this way and that until he eventually finds his way home. But random walks are not a very efficient means of getting anywhere: if the drunk's home is far away, he may well wake up the following morning in a bush on the other side of town. An object engaged in a random walk will tend to move away from its starting point by a distance proportional to the square root of the time taken. If in one minute a drunk has advanced by one metre, then after four minutes he will have advanced by two metres and after nine minutes, only three metres. Given this sluggish progress, it is not surprising that animals and microbes seldom use a random walk to find food or prey, only resorting to the strategy if no other options are available. Drop an ant onto unfamiliar ground and as soon as it encounters a scent, it will abandon a random walk and follow its nose.

Possessing neither nose nor navigation skills, the exciton energy was thought to advance through the chlorophyll forest via the drunkard's strategy. But such a picture didn't make much sense, as this first event in photosynthesis is known to be extraordinarily efficient. In fact, the transfer of captured photon energy from a chlorophyll antenna molecule to the reaction centre boasts the highest efficiency of any known natural or artificial reaction: close to 100 per cent. Under optimal conditions, nearly every energy parcel absorbed by a chlorophyll molecule makes it to the reaction centre. If the path taken were a meandering one, nearly all of them, certainly most of them, should get lost. How this photosynthetic energy can find its way to its destination so much better than drunkards, ants or indeed our most energy-efficient technology has been one of the biggest puzzles in biology.

The quantum beat

The senior author on the research paper[3] that sparked the newspaper article which had the MIT journal club laughing their quantum socks off was a naturalized American, Graham Fleming. Born in Barrow in the north of England in 1949, Fleming now heads a group at the University of Berkeley in California that is acknowledged as one of the world's leading research teams in this field, using a powerful technique with the impressive title of 'two-dimensional Fourier transform electronic spectroscopy' (2D-FTES). 2D-FTES can probe into the inner structure and dynamics of the tiniest molecular systems by targeting them with highly focused short-duration laser pulses. The group has performed most of its work studying not plants, but a photosynthetic complex called the Fenna–Matthews–Olson (FMO) protein that is made by photosynthetic microbes called green sulphur bacteria, found in the depths of sulphide-rich bodies of water such as the Black Sea. To probe the chlorophyll sample, the researchers fired three successive pulses of laser light into the photosynthetic complexes. These pulses deposit their energy in very rapid and precisely timed bursts and generate a light signal from the sample that is picked up by detectors.

Greg Engel, the lead author on the paper, spent the entire night stitching together the data generated from signals covering a time of 50 to 600 femtoseconds,* to produce a plot of their results. What he discovered was a rising and falling signal that oscillated for at least six hundred femtoseconds (see figure 4.7). These oscillations are akin to the interference pattern of light and dark fringes in the two-slit experiment; or the quantum equivalent of the pulsating sound beats heard when tuning a musical instrument. This 'quantum beat' showed that the exciton wasn't taking a single route through the chlorophyll maze but was instead following multiple routes simultaneously (figure 4.8). These alternative routes act a bit like the pulsed notes of the almost in-tune guitar: they generate beats when they are nearly the same length.

But remember that such quantum coherence is very delicate and extraordinarily difficult to maintain. Was it really feasible that a microbe or plant was able to beat the heroic efforts of the brightest and best of MIT quantum computing researchers to keep decoherence at bay? This was indeed the bold claim made in Fleming's paper, and it was this 'quantum hanky-panky', as Seth Lloyd described it,

---

* A femtosecond is one millionth of one billionth of a second, or 10^{-15} seconds.

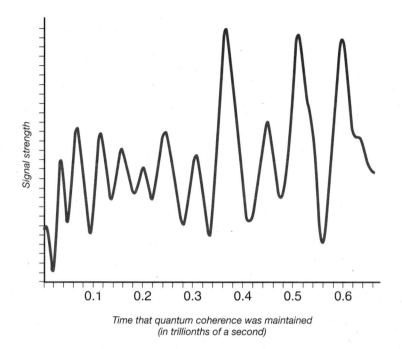

Signal strength

0.1   0.2   0.3   0.4   0.5   0.6

Time that quantum coherence was maintained
(in trillionths of a second)

Figure 4.7: The quantum beats seen by Graham Fleming and his colleagues in their 2007 experiment. What is important is not the irregular shape of the oscillations – it's the fact that there are oscillations at all.

that raised the hackles of the MIT journal club. The Berkeley group were suggesting that the FMO complex was acting as a quantum computer to find the quickest route to the reaction centre, a challenging optimization problem, equivalent to the famous travelling salesman problem in mathematics, which, for travel plans involving more than a handful of destinations, is solvable only with a very powerful computer.*

Despite their scepticism, the journal club set Seth Lloyd the task of investigating the claim. To everyone's surprise at MIT, the conclusion of Lloyd's scientific detective work was that there was indeed

---

*The travelling salesman's problem is to find the shortest route passing through a large number of cities. This is described mathematically as an *NP-hard problem*: that is, one for which no shortcut to a solution exists, even in theory, the only way to find the optimal solution being a computationally intensive, exhaustive search of all possible routes.

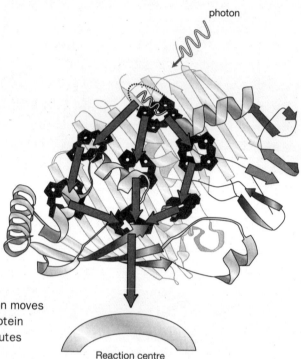

Figure 4.8: The exciton moves through the FMO protein following multiple routes at the same time.

photon

Reaction centre

substance to the Californian group's claims. The beats that Fleming's group had discovered in the FMO complex were indeed a signature of quantum coherence, and Lloyd concluded that the chlorophyll molecules were operating a novel search strategy known as a *quantum walk*.

The advantage of a quantum walk over a classical random walk can be appreciated by returning to our slow-moving drunk and imagining that the bar he leaves has sprung a leak and that water is pouring out of its door. Unlike our inebriated drinker, who must choose a single route, the waves of water escaping from the bar advance in all possible directions. Our drunken walker will soon find himself overtaken, as the watery wave advances through the streets at a rate simply proportional to the time taken, not its square root. So if at one second it had advanced by one metre, then after two seconds it will have covered two metres and after three seconds, three metres, and so on. Not only that, but because, like the superposed atom in the two-slit experiment, it travels by all possible routes simultaneously, some part of the wave front will definitely encounter the drunkard's home well in advance of the inebriated wanderer himself.

Fleming's paper caused its own wave of surprise and consternation that travelled well beyond the journal club at MIT. But some commentators were quick to point out that the experiments were conducted with isolated FMO complexes cooled to 77 K (a chilly −196°C): clearly far colder than any temperature compatible with plant photosynthesis or even life, but low enough to keep that pesky decoherence at bay. How relevant were these chilled bacteria to anything that goes on in the hot and messy interiors of plant cells?

It soon became clear, however, that quantum coherence was not limited to cold FMO complexes. In 2009, Ian Mercer at University College Dublin detected quantum beating in another bacterial photosynthetic system (or photosystem for short) called the Light Harvesting Complex II (LHC2), which is very similar to a plant photosystem, but at the normal ambient temperatures in which plants and microbes normally perform photosynthesis.[4] Then, in 2010, Greg Scholes of the University of Ontario demonstrated quantum beating in the photosystem of a group of aquatic algae (which, unlike the higher plants, lack roots, stems and leaves) called cryptophytes, which are extraordinarily abundant, to the extent that they are responsible for fixing as much atmospheric carbon (that is, extracting atmospheric carbon dioxide) as higher plants.[5] Around the same time, Greg Engel demonstrated quantum beating in the same FMO complex that he had studied in Graham Fleming's laboratory, but now at much higher, life-supporting, temperatures.[6] And just in case you might think this remarkable phenomenon is restricted to bacteria and algae, Tessa Calhoun and colleagues from Fleming's laboratory in Berkeley recently detected quantum beating in another LHC2 system, this time from spinach.[7] LHC2 is present in all higher plants and contains 50 per cent of all the chlorophyll on the planet.

Before moving on we will describe briefly just how the solar-derived exciton energy is used, as Feynman described, to knock 'this oxygen away from the carbon . . . leaving the carbon, and water, to make the substance of the tree' – or apple.

After enough energy arrives at the reaction centre, it causes a special pair of chlorophyll molecules (called P680) to spit out electrons. We will be learning a bit more about what goes on within the reaction centre in chapter 10, as it is a fascinating place that may house another novel quantum process. The source of these electrons is water (which, remember, is one of the ingredients in Feynman's description of photosynthesis). As we discovered in the last chapter, the capturing of electrons from any substance is called oxidation, and it is the same process that takes place during burning. When wood burns in air, for example, oxygen atoms pull electrons from carbon atoms. The electrons in the outer orbit of carbon are fairly loosely attached, which is why carbon burns easily. However, in water they are held very tightly: photosynthesis systems are unique in that they are the only place in the natural world where *water* is 'burnt' to yield electrons.*

So far, so good: we now have a supply of free electrons thanks to the energy delivered by the excitons in chlorophyll. Next, the plant needs to send these electrons where they can be put to work. They are first captured by the cell's designated electron transporter, NADPH. We met a similar molecule, NADH, briefly in the last chapter, where it was involved in ferrying electrons captured from nutrients, such as sugars, to the respiratory chain of enzymes in the cell's energy organelles, the mitochondria. If you remember, the captured electrons delivered to the mitochondria by NADH then flowed down a respiratory chain of enzymes as a kind of electric current that is used to pump protons across a membrane, and the resulting backflow of these protons is used to make the cell's energy carrier, ATP. A very similar

---

* When we say the 'burning of water' we of course do not mean that water is a fuel like coal, but rather we are using the term loosely to denote the molecular process of oxidation.

process is used to make ATP in plant chloroplasts. NADPH feeds the electrons it is carrying into a chain of enzymes that similarly pumps protons out of the chloroplast membrane. The backward flow of these protons is used to produce ATP molecules, which can then go on to power lots of energy-hungry processes in the plant cell.

But the actual carbon fixation process, the capture of carbon atoms from carbon dioxide in air and their use to make energy-rich organic molecules like sugars, takes place outside the thylakoid, though still within the chloroplast. This is performed by a big bulky enzyme called RuBisCO that is probably the most abundant protein on earth as it has to do the biggest job: making nearly all the world's biomass. This enzyme bolts the carbon atom pulled from carbon dioxide onto a simple five-carbon sugar molecule called ribulose-1,5-bisphosphate to make a six-carbon sugar. To achieve this feat it makes use of the two ingredients it has been supplied with: electrons (delivered by NADPH) and a source of energy (ATP). Both ingredients are the products of the light-driven processes of photosynthesis.

The six-carbon sugar made by RuBisCO immediately breaks down into two three-carbon sugars which are then bolted together in lots of different ways to make all of the biomolecules that make an apple tree, including its apples. The inanimate New England air and water have, with the help of light and a sprinkling of quantum mechanics, become the living tissue of a New England tree.

By comparing photosynthesis in plants with the respiration (burning our food) that takes place in our own cells, discussed in the last chapter, you can see that, under the skin, animals and plants are not so different. The essential distinction lies in where we, and they, get the fundamental building blocks of life. Both need carbon, but plants obtain it from air whereas we get it from organic sources, such as the plants themselves. Both need electrons to build biomolecules: we *burn* organic molecules to capture their electrons, while plants use light to *burn* water to capture its electrons. And both need energy: we scavenge it from the high-energy electrons that we obtain from our food by running them down respiratory energy hillsides; plants capture the energy of solar photons. Each of these processes involves the motion of fundamental particles that are governed by quantum rules. Life seems to be harnessing quantum processes to help it along.

The discovery of quantum coherence in warm, wet, turbulent systems such as plants and microbes has come as a huge shock to quantum physicists, and a great deal of research is now focused on working out precisely how living systems protect, and utilize, their delicate quantum coherent states. We will be returning to this puzzle in chapter 10, where we will examine some very surprising possible answers that may even help physicists, such as those MIT quantum theorists, to build practical quantum computers that could work on your desk, rather than in your deep freeze. The research is also likely to inspire a new generation of artificial photosynthetic technologies. Current solar cells are loosely based on photosynthetic principles and are already competing with solar panels for a share of the clean energy market, but their efficiency is limited by losses during energy transport (at best about 70 per cent efficiency compared to the near 100 per cent efficiency of the photon energy capture step in photosynthesis). Bringing biology-inspired quantum coherence to solar cells has the potential to greatly increase the efficiency of solar energy and thereby deliver a cleaner world.

Finally in this chapter, then, let us take a moment to consider the significance of what we have added to our understanding of what is special about life. Consider again those quantum beats that Greg Engel first saw in his FMO complex data, which show that particles move within living cells as waves. There is a temptation to think of these as laboratory-bound phenomena with no significance outside biochemical experimentation. But subsequent research has demonstrated that they do indeed exist in the natural world too, inside leaves, algae, and microbes, and that they play a role, possibly a crucial one, in building our biosphere.

Still, the quantum world appears very strange to us and it is often claimed that this strangeness is a symptom of a fundamental split between the world we see around us and its quantum underpinnings. But in reality there is only a single set of laws that govern the way the world behaves: quantum laws.* The familiar statistical laws and Newtonian laws are, ultimately, quantum laws that have been filtered through a decoherence lens that screens out the weird stuff (which is why quantum phenomena appear weird to us). Dig deeper and you will always find quantum mechanics lurking at the heart of our familiar reality.

What's more, certain macroscopic objects *are* sensitive to quantum phenomena; and most of these are living. We discovered in the last chapter how quantum tunnelling inside enzymes can make a difference to whole cells; and here we have explored how the initial photon-capturing event responsible for putting most of the biomass on the planet appears to be dependent on a delicate quantum coherence that can be maintained for biologically relevant lengths of time within the warm but highly organized interior of a leaf or microbe. Once again we see Schrödinger's *order from order* capable of capturing quantum events, and what Jordan termed *amplification* of quantum phenomena into the macroscopic world. Life seems to bridge the quantum and classical worlds, perched on the quantum edge.

We next turn our attention to another essential process for our biosphere. Newton's apple tree wouldn't have been able to make any apples if its blossom hadn't first been pollinated by birds and insects, particularly bees. But the bees had to find the apple blossom; and they did so using another capability believed by many to be driven by quantum mechanics – the sense of smell.

---

* We should add a qualification here since quantum mechanics cannot so far account for the gravitation force, as general relativity (which is how we understand gravity) appears to be incompatible with quantum mechanics. Unifying quantum mechanics and general relativity to construct a quantum theory of gravity remains one of the greatest challenges confronting physics.

5

Finding Nemo's home

The nose, for example, of which no philosopher has ever spoken with veneration and gratitude – the nose is, albeit provisionally, the most delicate instrument at our disposal. It is an instrument capable of recording the most minimal changes that escape even spectroscopic detection.

Friedrich Nietzsche, *Twilight of the Idols*, 1889

They appear to be bringing us a certain message from a material reality.

Gaston Bachelard, *La Formation de l'esprit scientifique, contribution à une psychoanalyse de la connaissance objective*, 1938

Tucked within the arms of a deadly sea anemone that is fastened to a coral reef off the coast of Isla Verde in the Philippines is a pair of small orange-and-white striped fish known as common clownfish or, more properly, anemonefish, or, even more properly, *Amphiprion ocellaris*. One of the pair, a female, has led a more interesting life than most vertebrates, because she has not always been female. Like all anemonefish, she started life as a smaller male who had been subordinate to the one female in the group of fish inhabiting this particular anemone. Anemonefish have a rigid social structure, and as a male this one had competed with the other males until eventually becoming dominant and enjoying the honour of mating with the sole female. But when its mate was eaten by a passing eel, the ovaries that had lain dormant in its body for several years matured, its testes ceased to function and the male clownfish became the queen female ready to mate with the next male in the pecking order.

Anemonefish are common inhabitants of coral reefs from the Indian Ocean to the western Pacific, feeding on plants, algae, plankton and animals such as molluscs and small crustaceans. Their small size, bright colours and absence of spines, sharp fins, barbs or spikes make them easy prey for the eels, sharks and other predators that rove the reef. When threatened, their principal means of defence is to dash between the tentacles of their host anemone, from whose poisonous sting they are protected by a thick layer of mucus covering their scales. In turn, the anemone benefits from its colourful tenants who chase off unwelcome intruders, such as grazing butterflyfish.

It was in this setting that the anemonefish came to be most familiar to us in the animated film *Finding Nemo*.* The challenge facing Nemo's dad, Marlin, was to find his son, who had been abducted from his home in the Great Barrier Reef and carried all the way to Sydney. But the challenge that besets real anemonefish is to find their way back home.

Each anemone may be host to an entire colony of anemonefish that contains a dominant male and female together with several juvenile males vying with each other for the role of queen's consort. The unusual capability of the dominant male to change sex on the death of the queen fish, a capability known as protandrous hermaphroditism, may be an adaptation to life in the dangerous reef, as it allows the colony to survive the demise of the single reproductive female without ever having to leave the protection of the host anemone. But although an entire colony of fish may remain resident on a

---

* Sadly, that popularity now threatens the animal in the wild, as it has become a favourite of poachers who over-collect the fish to feed the burgeoning aquarium market in anemonefish. So don't keep Nemo in your home; he or she belongs on real coral reefs!

single anemone for many years, the progeny of those fish must leave the safety of their home. And, eventually, they will need to find their way back.

A full moon is the cue for spawning of most coral fish.* As the moon begins to wane over the ocean, the female of the pair busies herself laying a clutch of eggs to be fertilized by the dominant male. Thereafter, her work is done; guarding the eggs and chasing away carnivorous reef fish is the job of the male anemonefish. After about a week of his custodianship, the eggs hatch and hundreds of larvae are launched into the currents.

Larval anemonefish are only a few millimetres in length and almost completely transparent. For about a week they drift in the pelagic currents, feeding on zooplankton. As anyone who has dived off coral reefs will know, drifting in an ocean current will soon take you far from your starting point; so anemonefish larvae can be carried many kilometres from their natal reef. Most are eaten, but some survive; after about a week, these lucky few swim to the sea floor and, within a day, metamorphose (like our frog in chapter 3) into their juvenile form, a smaller version of the adult fish. Lacking the protection of the poisonous anemone, the brightly coloured juvenile is very vulnerable to predators that cruise the benthic waters. If it is to survive, it must quickly find a coral reef where it can gain sanctuary.

It was thought that larval reef fish drifted with the ocean currents and that they relied on mere chance to be washed up close to a suitable reef. But that explanation didn't really make complete sense since it was known that most larvae are strong swimmers and there is no point in swimming if you don't know where to go. Then in 2006 Gabriele Gerlach, a researcher from the famous Marine Biological Laboratory in Woods Hole, Massachusetts, carried out genetic fingerprinting of fish living on reefs separated by between 3 and 23 kilometres within the complex that forms the Great Barrier Reef of Australia. She discovered that fish inhabiting the same reef were

---

* The stronger tides at this time are thought to aid dispersal.

much more closely related to each other than they were to those inhabiting more distant reefs. Since all juvenile reef fish larvae disperse across large distances, the finding only makes sense if most adults return to the reef on which they were born. Somehow, each larval reef fish must be imprinted with a signature that identifies its spawning area.

But how do larval or juvenile anemonefish that have drifted so far from their home know which direction to swim in? The sea floor doesn't provide any useful visual cues. It lacks reference points and so looks the same in all directions: a sandy desert decorated with scattered pebbles, boulders and the occasional wandering arthropod. The distant coral reef is unlikely to provide any auditory signals that could travel several kilometres. The currents themselves are an additional problem, as the direction of flow varies with depth and it can be very difficult to determine whether the body of water is moving or stationary. There is no evidence that anemonefish possess the kind of magnetic compass that helped to guide our robin on her winter migration. So how do they find their way?

Fish do possess a keen sense of smell. Sharks, two-thirds of whose brains is devoted to olfaction, can famously smell a drop of blood from more than a kilometre away. Perhaps reef fish smell their way home? To test this theory, in 2007 Gabriele Gerlach designed a 'two-channel olfactory choice flumes test' in which larval reef fish were placed downstream of two flumes of seawater: one collected from the reef on which they hatched and the other from a distant reef. She then measured which water flume the larvae preferred: home or away?

Invariably, the larval fish swam towards the flume filled with water from the reef on which they had been spawned. They could clearly discriminate between foreign and native reef waters, presumably by their different smells. Michael Arvedlund, a researcher from James Cook University in Queensland, Australia, used a similar experimental setup to demonstrate that anemonefish could smell their host anemone species and distinguish it from others that they do not colonize. Even more remarkably, Daniella Dixson, also from James Cook University, found that anemonefish can distinguish between water collected from their preferred habitat of reefs that lie beneath vegetated islands and the less favoured offshore reef water. It really seems that Nemo and other reef fish sniff their way home.

The ability of animals to navigate with their sense of smell is legendary. Every year, along ocean coasts around the world, millions of salmon assemble into large schools at the mouths of rivers before venturing inland to battle against the flow of current, rapids, water-falls and sandbanks to reach their spawning grounds. As with the anemonefish, it was thought that the salmon's selection of a suitable river was pretty much down to chance. But then in 1939 the Canadian Wilbert A. Clemens tagged 469,326 young salmon caught in a particular tributary of the Fraser River system. Years later, he caught 10,958 of those he had tagged that had returned to that same tributary. Not a single marked salmon was caught in any other tributary of the river. None had lost its way on its journey from the ocean to their home stream. How they manage to navigate through ocean and stream remained a mystery for many years. Then Professor Arthur Hasler from the University of Wisconsin-Madison suggested that the young salmon follow a scent trail, and tested his theory in 1954 by catching several hundred returning fish upstream of a fork (the confluence of two streams) in the Issaquah River near Seattle and then transporting them downstream to below the fork. The salmon invariably returned to the same branch of the fork in which they had been captured. But when he blocked their nostrils with cotton-wool stoppers before release, they swam up to the fork in the river tacking this way and that and could not decide whether to go right or left.

The olfactory sense is perhaps even more remarkable on land because the volume of the atmosphere, in which odorants are diluted, is even vaster than that of the ocean. The atmosphere is also subject to a greater degree of turbulence, owing to the weather, so odorant molecules are dispersed more quickly in air than in water. Yet the sense of smell is vital to the survival of most land animals, used not only to find the way home but also to catch prey, escape from predators, find a mate, provide alarm signals, mark territory, trigger physiological changes and communicate. This whole *smellscape* of olfaction is much less obvious to humans, who often harness the keener olfactory senses of their companion animals to detect these signals and signs. Dogs are, of course, famously interested in smell and the bloodhound, whose olfactory epithelium (more on this later) is forty times the size of ours, is rightly famous for its ability to follow the scent trail of a single individual. We have all seen those movies featuring a keen tracker dog that needs only a quick sniff of the discarded shirt of an escaped convict to be able to track the villain across moorland, forest and stream. And though the stories may be fiction, the ability of the hound is entirely real. Dogs can tell from a track which way the person or animal was travelling and can follow a scent trail that is several days old.

The startling power of animals' sense of smell can be appreciated if we reflect upon the feats that a bloodhound or an anemonefish routinely accomplishes. Consider first the bloodhound; its sense of smell is tuned to detect tiny quantities of organic chemicals, such as butyric acid, that are shed by humans and other animals; and the sensitivity of its nose is extraordinary. If just a single gram of butyric acid were allowed to evaporate in a room, then we humans would be just about capable of detecting its sweet, rancid odour. But a dog is able to detect that same gram of chemical if its vapours were diluted to fill the air above an entire city to a height of 100 metres. And consider again those anemonefish or salmon that detect the scent of their distant home, kilometres away, diluted in the vastness of the ocean.

But the animal sense of smell isn't remarkable only for its sensitivity. There is also its highly developed discriminatory power. Dogs are routinely used by customs officers to detect a wide range of odorants, from drugs such as marijuana and cocaine, to chemicals in explosives such as C-4 – often through dense packaging and a suitcase. They can also distinguish between the scents of individuals, even identical twins. So how do they do this? Surely the butyric acid shed by one of us is the same butyric acid that is released by everyone? Of course it is; but, alongside the butyric acid, each of us sheds a delicate and complex cocktail of hundreds of organic molecules that provides a signature of our presence that is as individual as our fingerprints. Dogs can 'see' our olfactory fingerprint as easily as we can see the colour of a person's shirt. Anemonefish or salmon must similarly recognize the scent of their home, just as we might recognize our street or spot the colour of our front door.

But dogs, salmon or anemonefish aren't the supreme athletes of olfaction. A bear's sense of smell is over seven times as sensitive as even a bloodhound's; and it can smell a carcass 20 kilometres away. A moth can detect a mate at a distance of some 10 kilometres; rats smell in stereo and snakes smell with their tongues. All these olfactory skills are essential for animals that must seek out food, find mates and/or avoid predators; they have evolved a sensitivity to volatile cues that betray the proximity of these resources or dangers, whether in air or water. The sense of smell is so important to animal survival that behavioural responses to odours appear to be hardwired in a number of species. Experiments with Orkney Island voles demonstrated that they avoided traps baited with the secretions of predatory stoats, even though stoats have been absent from the island for five thousand years!

Humans are said to have a much poorer sense of smell than our relatives. When, several million years ago, *Homo erectus* lifted his upper body off the forest floor to walk upright, he also raised his nose from the ground and its rich source of aromas. Thereafter, sight and sound, both more efficient from a higher vantage point, became his principal sources of information. So the human snout became shorter, the nostrils narrowed and mutations accumulated in most of the thousand or so ancestral mammalian genes that encode olfactory receptors (more on these later). We also, perhaps sadly, lost an auxiliary olfactory sense found in other animals and conferred by the vomeronasal organ (VNO) or Jacobson's organ, whose role is to detect sex pheromones.

Yet, despite our diminished genetic repertoire of only about three hundred olfactory receptor genes and our altered anatomy, we have retained a surprisingly good sense of smell. We may not be able to sniff out a mate or our dinner from several miles away, but we can discriminate between around ten thousand different scents and, as Nietzsche noted, can outperform 'even spectroscopic detection' of odorous chemicals. Our ability to appreciate scents has inspired some of our greatest poetry ('A rose by any other name would smell as sweet') and plays a crucial role in our sense of well-being and contentment.

Our sense of smell has also played a surprisingly active role in human history. The earliest texts record a reverence for pleasant aromas and an abhorrence of foul smells. Places of worship and meditation were frequently scented with perfumes and spices. In the Hebrew Bible, God instructs Moses to build a place of worship and tells him: 'Take to you sweet spices, stacte, and onycha, and galbanum; these sweet spices with pure frankincense: of each shall there be a like weight. And you shall make it a perfume, a confection after the art of the apothecary, tempered together, pure and holy.'[1] The ancient Egyptians even had a god of perfume, Nefertum, who was also a god of healing, a kind of mythical aromatherapist.

The association of health with pleasant aromas, and, conversely, disease and decay with foul smells, led many to believe that the causal direction led from odour to health or disease, rather than the reverse. For example, the great Roman physician Galen taught that malodorous sheets, mattresses and blankets could accelerate the pollution of the body fluids. Nauseating exudations (miasmas) coming from sewers, charnel houses, cesspools and marshes were considered to be the sources of many fatal diseases. Conversely, pleasant smells were thought to ward off illness so that, in medieval Europe, physicians would insist that, before they entered a plague victim's house, it had to be thoroughly aired and perfumed by lighting fragrant fires scented with incense, myrrh, roses, cloves and other aromatic herbs. Indeed, the profession of perfumery was originally dedicated to the disinfection of houses, rather than personal grooming.

The importance of the sense of smell is not of course limited to detecting odorants breathed in through our nostrils. Remarkably, our sense of *taste* is generally considered to be about 90 per cent smell. When we taste food, the taste receptors on our tongue and palate detect chemicals dissolved in saliva; but the receptors come in only five varieties, able to identify combinations of only five basic tastes – sweet, sour, salty, bitter and umami (a Japanese word that means 'pleasant savoury taste'). But volatile odorants evaporating from our food and drink gain entry to the nasal cavity from the back of our throat to activate combinations of hundreds of different smell receptors. These provide us with a far greater ability, compared with taste, to distinguish between thousands of different aromas and to enjoy the rich *flavours* (mostly scents) of fine wine, aromatic food, spices, herbs or coffee. And even though we have lost the vomeronasal sense enjoyed by most of our fellow mammals, the huge perfume industry is evidence of the role that scent continues to play in human courtship and sex. Freud even saw a connection between sexual repression and the sublimation of the sense of smell in most of us, but nevertheless claimed that 'there exist, even in Europe,* peoples who are highly appreciative of the strong odour of the genitalia.'[2]

So how do humans, dogs, bears, snakes, moths, sharks, rats or anemonefish detect these messages 'from a material reality'? How do we distinguish between such a wide variety of odorants?

The physical reality of odours

Unlike our senses of sight and hearing, which capture information indirectly via electromagnetic waves or sound waves carried to us from an object, both taste and smell receive information directly

* Note the implicit racism.

from contact with the object detected (a molecule), bringing messages 'from a material reality'. Both appear to work through rather similar principles. The molecules they detect are either dissolved in saliva or float through the air and are then picked up by receptors either on the tongue (taste) or in the olfactory epithelium in the roof of our nasal cavity (smell). This requirement for volatility means that most odorants are fairly small molecules.

The nose itself plays no direct role in smelling, other than channelling air towards the olfactory epithelium, which is at the back of the nose (figure 5.1). This tissue is quite small, measuring only 3 square centimetres (about the size of a postage stamp) in humans, but it is lined with both mucus-secreting glands and millions of *olfactory neurons*. These are a type of nerve cell that are to the sense of smell what retinal rods and cones in the eye are to the sense of sight. The front end of the olfactory neuron is shaped a bit like a broom, with a many-pronged head where the cell membrane is folded into lots of hair-like *cilia*. This broom with its brush of cilia pokes out of the cell layer where it can capture passing odour molecules. The back end of the cell is like the broom's handle, forming

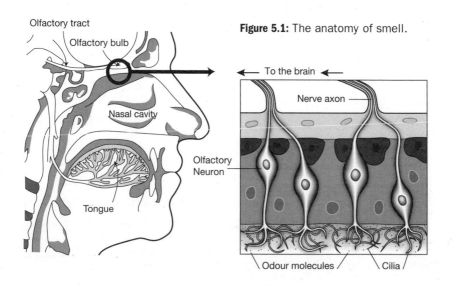

Figure 5.1: The anatomy of smell.

the cell's *axon* or nerve that extends through a small bone at the back of the nasal cavity to enter the brain, where it connects with a region called the *olfactory bulb*.

The remainder of this chapter may be best read with an orange in front of you, perhaps chopped into segments so that the tangy aromas are released and travel through your nose to reach the nasal epithelium. You might even slip one of the segments into your mouth to allow its volatile odorants to find their way through the retronasal route to that same tissue. Like all natural scents, the smell of an orange is very complex and made up of hundreds of volatile compounds, but one of the most fragrant is called limonene,* whose course we will now follow from molecule to fragrance.

Limonene, as its name implies, is abundant in citrus fruits such as oranges and lemons and is largely responsible for their tangy aroma and flavour. The chemical belongs to a class of compounds known as *terpenes*, which are the smelly constituents of the essential oils of many plants and flowers, generating the rich aromas of pine, roses, grapes and hops. So feel free to exchange the orange for a glass of beer or wine if you prefer. This chemical is produced in many parts of the citrus plant, including its leaves, but is most abundant in the skin of its fruit, which can be squeezed to yield nearly pure limonene.

Limonene is a volatile liquid that will gradually evaporate at room temperature, so your orange will be releasing millions of limonene molecules into the surrounding air. Most of these will float off into the room and out through the door and windows, but a few will be carried close to your nose by the air currents. Your next inhalation will suck in several litres of this odorant-laden air that will pass through your nostrils and across the nasal epithelium, which is lined with approximately ten million olfactory neurons.

When limonene molecules waft past the olfactory epithelium brushes, some of them are ensnared by the olfactory neurons. The

---

* 1-methyl-4-(1-methylethenyl)-cyclohexene.

capturing of a single limonene molecule is sufficient to trigger the opening of a tiny channel in the neuron's cell membrane that allows a flow of positively charged calcium ions into the cell from outside. When about thirty-five limonene molecules have been captured, the subsequent flow of ions into the cell amounts to a tiny electric current of about one picoamp* in total. This level of current acts like a switch to fire an electrical signal, called an action potential (we will learn much more about these in chapter 8) down the handle of the brush-like cell, its axon. This signal travels all the way to the olfactory bulb in your brain. After further neural processing you experience this 'message from a material reality' as the tangy aroma of oranges.

The key event in this whole process is of course the capturing of the odour molecule by the olfactory neuron. So how does it work? By analogy with sight and the eye's light-sensitive cone and rod cells (also types of neurons), it was expected that the sense of smell would similarly be conferred by some kind of surface-located olfactory receptors. But, in the 1970s, the nature and identity of olfactory receptors were completely unknown.

Richard Axel was born in Brooklyn, New York, in 1948, the first child of immigrant parents who had fled Poland before the Nazi invasion. His childhood was typical for the neighbourhood: running errands for his father, a tailor, between playing stickball (a kind of street baseball, with manhole covers as the bases and a broom handle as bat) or basketball in the local roads and courtyards. His first job, aged eleven, was as a messenger, delivering false teeth to dentists; at twelve he was laying carpets, and at thirteen serving corned beef and pastrami in a local delicatessen. The chef was a Russian who used to recite Shakespeare while slicing cabbage heads, providing the young Richard with his first real exposure to the cultural world beyond delis and basketball courts and inspiring a deep and abiding love of great literature. Axel's intellectual talents were spotted by a local

---

* A picoamp is one trillionth (10^{-12}) of an amp.

high-school teacher who encouraged him to apply, successfully, for a scholarship to Columbia University in New York to read literature.

As a freshman, Axel threw himself into the intellectual maelstrom of university life in the 1960s. But to support his party-going lifestyle he took a job washing glassware in a molecular genetics laboratory. He became fascinated by this emerging science, but remained hopeless at glass washing, so was sacked from that job and rehired as a research assistant. Torn between literature and science, he eventually decided to enrol in a graduate genetics course but then switched to studying medicine to escape the Vietnam draft. He was apparently as bad at medicine as he'd been at glass washing. He couldn't hear a heart murmur and never saw the retina; his glasses once fell into an abdominal incision and he even managed to sew a surgeon's finger to his patient. He was eventually allowed to graduate only on condition that he promise never to practise medicine on living patients. He returned to Columbia to study pathology, but after a year the chairman of the department insisted that he should never practise on dead patients either.

Realizing that medicine clearly lay beyond his talents, Axel eventually managed to return to research at Columbia University. Thereafter he made rapid progress and even invented a novel technique for getting foreign DNA inside mammalian cells that became a mainstay of the genetic engineering/biotech revolution of the late twentieth century and earned Columbia University hundreds of millions of dollars of revenue in licensing agreements: a generous return on their scholarship investment.

By the 1980s Axel was wondering whether molecular biology could help to solve that mystery of mysteries: how the human brain works. He switched from studying the behaviour of genes to studying the genes for behaviour, with the long-term aim of 'dissecting how higher brain centers generate a "percept," say, of the scent of a lilac, or coffee, or a skunk . . . '.[3] His first foray into neuroscience was investigating egg-laying behaviour in a marine snail. It was at about this time that a very talented researcher, Linda Buck, joined his lab.

She had trained as an immunologist at the University of Dallas before becoming fascinated by the emerging field of molecular neuroscience and moving to Axel's laboratory to be at the forefront of this research. Together, Axel and Buck devised an ingenious series of experiments to probe the molecular basis of smell. The first question they addressed was the identity of the receptor molecules that were presumed to exist on the surface of olfactory neurons and to capture and identify different odorant molecules. Extrapolating from what was known about other sensory cells, they guessed that the receptors were some kind of proteins poking out of the cell membrane where they could bind passing odour molecules; but, at the time, nobody had ever isolated any of these odour receptors so no one had a clue what they looked like or how they worked. All the team had to go on was an inkling that the elusive receptors might belong to a family of proteins called G-protein-coupled receptors that were known to be involved in detecting other kinds of chemical signals, such as hormones.

Linda Buck managed to identify an entirely new family of genes encoding this kind of receptor that were only expressed* in olfactory receptor neurons. She went on to demonstrate that these genes did indeed encode the elusive odour-capturing receptors. Further analysis showed that the rat's genome encoded about a thousand of these newly identified receptors, each a little different from the others, and each presumably tuned to detect a single odorant. Humans have a similar number of olfactory receptor genes, but two-thirds of them have degenerated into what are called *pseudogenes*, which are a kind of gene fossil that have accumulated so many mutations that they no longer work.

But whether there are three hundred or a thousand receptor genes, this is very far from the figure of ten thousand different scents that humans can identify. There clearly wasn't a one-to-one mapping

---

*In this context 'expressed' means a gene that is active in the sense that its information is copied into RNA, which then feeds into the protein synthesis machinery to make the protein encoded by that gene, such as an enzyme or particular olfactory receptor.

between types of odour receptors and types of odours; so how the signals received by olfactory receptors are transformed into smells remained a mystery. It also wasn't clear how the job of detecting all the variety of odour molecules was shared out between different cells. The genome of each cell has the complete set of olfactory receptor genes, so could potentially detect every odour. Or is there some kind of division of labour? To answer these questions the Columbia University team devised an even more ingenious experiment. They genetically altered mice so that all olfactory neurons expressing one particular odour receptor were dyed blue. If all the cells stained blue, that would indicate that they all expressed this receptor. The answer was clear when the team examined the olfactory cells of the engineered mice: approximately one in every thousand cells was dyed blue. It seemed that each olfactory neuron was not a generalist but a specialist.

It wasn't long before Linda Buck moved from Columbia to set up her own laboratory at Harvard, and the two groups continued working in parallel to dissect many of the remaining secrets of olfaction. They soon devised techniques to isolate individual olfactory neurons and directly probe their sensitivity to particular odorants, such as the limonene of your orange. They discovered that each odorant chemical activated not just one but several neurons; also, that single neurons responded to several different odorants. These findings appeared to solve the conundrum of how only three hundred olfactory receptors can identify ten thousand different smells. Just as only twenty-six letters can be combined in lots of different ways to write every word in this book, so a few hundred olfactory receptors can be activated in trillions of different combinations to provide the vast array of scents.

Richard Axel and Linda Buck were awarded the Nobel Prize in 2004 for their pioneering discoveries of 'odorant receptors and the organization of the olfactory system'.

Unlocking the odour key

The initiating event in the detection of an odour, such as that of an orange, a coral reef, a mate, a predator or prey is now understood to be the binding of a single molecule of odorant to a single olfactory receptor on the surface of the brush end of one of those broom-like olfactory neurons. But how does each receptor recognize its own set of odorant molecules, such as limonene, and not capture and bind to any one of the chemical ocean of other possible odorants that might float past the olfactory epithelium?

This is the central mystery of smell.

The conventional explanation is based on what is known as the lock-and-key mechanism. The odour molecules are thought to fit into the olfactory receptors like a key in a keyhole. For example, the limonene molecule was thought to slip snugly into a specialist olfactory receptor. Somehow, in a process that remains unclear, this binding event was thought to turn the lock of the receptor and trigger the release of a protein, called a G protein, that is normally tethered to the inner surface of the receptor, rather like a torpedo tethered to the hull of a ship. Once the torpedo protein is fired into the cell it makes its way to the cell membrane where it opens a channel that allows electrically charged molecules to flow into the cell. This electric current flowing through the membrane triggers the neuron to fire (more on this in chapter 9) and send a nerve signal that travels all the way from the olfactory epithelium to the brain.

The lock-and-key mechanism proposes that the receptor molecules are complementary in shape to the odour molecules, which fit inside them. A simple analogy is the shape-fitting puzzles that toddlers enjoy, in which a block cut into a particular shape (say, a circle, square or triangle) has to fit into a wooden board with the complementary shape cut out. We can think of each odour molecule as one of the block shapes – so that, perhaps, an orange odorant such as limonene is a circle, an apple odorant is a square, and a banana odorant a triangle. We can then imagine each olfactory receptor as possessing an *odorant binding pocket* that is moulded into the ideal shape for the olfactant molecule to fit neatly inside.

Of course, real molecules rarely come in such neat shapes, so real receptor proteins are presumed to have much more complex binding pockets to fit the more intricate shapes of real odorant molecules. Most are probably highly complex shapes similar to the active sites of enzymes that, as you will recall from chapter 3, bind substrate molecules. Indeed, odorant molecules are believed to interact with binding pockets in a fashion akin to the way that enzyme substrates are tethered into the active sites of enzymes (figure 3.4), or even the way that drugs interact with enzymes. Indeed, it has been argued that understanding the role that quantum mechanics plays in the interaction between olfactants and their receptors could eventually lead to more efficient drug design.

In any event, a clear prediction of the shape theory is that there should be some kind of correlation between the molecular shape of an odorant and its smell: similarly shaped odorant molecules should smell alike and very differently shaped molecules are likely to have sharply distinct odours.

One of the most feared scents in human history was the smell of mustard or rotten hay in the trenches of the Great War. Invisible gases would float across no-man's-land and just the faintest whiff of mustard (mustard gas) or musty hay (phosgene) might be sufficient to give a soldier a few precious seconds in which to don his mask before the deadly substance filled his lungs. The chemist Malcolm Dyson survived a mustard gas attack, and maybe it was this insight into the survival value of a keen nose that led him to ponder the nature of scent, because after the war he went on to synthesize many industrial compounds and to use his nose to smell the products of his synthetic reactions. But Dyson was puzzled by the apparent absence of any obvious relationship between the shape of a molecule and its smell. For example, many molecules that have very different shapes, such as the compounds in figure 5.2 **a–d**, smell the same – in

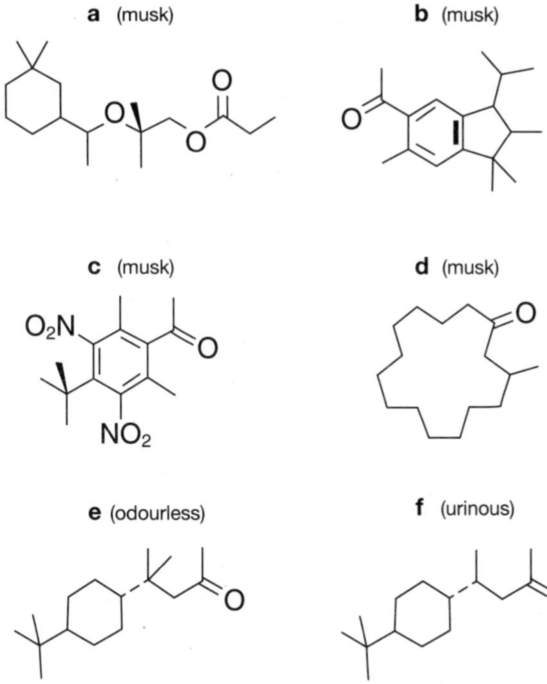

Figure 5.2:
Molecules (a)–(d) have very different shapes but smell pretty much the same. Molecules (e) and (f) have nearly identical shapes but have very different smells.

this case, they all smell musky.* Conversely, compounds that have very similar structures (such as compounds **e** and **f** in the figure) often have very different smells – in this case compound **f** smells like urine whereas **e** has no smell at all.[4]

This far from straightforward connection between the shape of a molecule and its odour was, and still is, a major problem for the industrial manufacturers of perfumes, flavours and fragrances. Instead of being able to design a perfume in the way they might design the shape of the perfume bottle, perfumers are forced to rely on chemical synthesis by brute force and trial-and-error sniff tests by chemists such as Dyson. But Dyson noticed that odour groups (chemicals that smell the same) were often composed of compounds

---

*Traditionally, musk was obtained from a number of natural sources, including the sex glands of the musk deer, the face glands of the musk ox, the faeces of the pine marten and the urine of the rock badger. But nearly all perfume musk is now synthetic.

that incorporated the same chemical groups, for example, the oxygen atom linked to a carbon atom by a double C=O bond in the musky-smelling chemicals in figure 5.2. These chemical groups are the component parts of any large molecule and determine many of its properties, apparently including, as Dyson noted, its scent. Another set of compounds with a similar smell is the large number of chemicals, with diverse molecular shapes, that possess a *hydrogen sulphide* (S–H) group, in which a hydrogen atom is attached to a sulphur atom, and that have the characteristic rotten-egg smell. Dyson went on to propose that what the nose detects is not the shape of an entire molecule but rather a different physical feature, namely the frequency at which the molecular bonds between its atoms vibrate.

In the late 1920s, when Dyson first made these claims, no one had any idea how to detect molecular vibrations. But on a voyage to Europe in the early 1920s the Indian physicist Chandrasekhara Venkata Raman was enchanted by 'the wonderful blue opalescence of the Mediterranean sea' and speculated that 'the phenomenon owed its origin to the scattering of light by molecules of water'. Normally, when light bounces off an atom or molecule it does so 'elastically', that is, without losing any energy, rather like a hard rubber ball bouncing off a rigid surface. Raman suggested that on rare occasions light can scatter 'inelastically', rather like a hard ball hitting a wooden bat and transferring some of its energy into the bat and the batsman (think of Bugs Bunny whacking a fast baseball so hard he sets both bat and bunny vibrating). In inelastic scattering, photons similarly lose energy to the molecular bonds they bump into, causing them to vibrate; the scattered light therefore emerges with less energy. Reducing the energy of light lowers its frequency and shifts its colour towards the blue end of the spectrum, providing Raman with his 'wonderful blue opalescence'.

Chemists utilize this principle to probe molecular structure. Essentially, light is shone on a chemical sample and the difference in colour or frequency (hence energy) between the input and output light is recorded as a *Raman spectrum* for a particular chemical, which provides a kind of signature of its chemical bonds. The technique bears its inventor's name, Raman spectroscopy, and earned him a Nobel Prize. When Dyson heard about Raman's work he saw that it could provide a mechanism by which the nose might probe the molecular vibrations of odour molecules. He proposed that the nose 'may be a spectroscope' capable of detecting the signature frequencies at which different chemical bonds vibrate. He even identified common frequencies in the Raman spectra of compounds that correlated with their odours. For example, all *mercaptans* (compounds that contain a terminal sulphur–hydrogen bond) share a particular Raman peak with frequency of 2567–2580. And they all stink of rotten eggs.

Dyson's theory did at least account for the analytical nature of odours, but no one had the slightest idea how anything like Raman spectroscopy could be harnessed by our nose to provide the sense of smell. After all, not only would the scattered light need to be captured and analysed by any biological spectroscope, there would also need to be a source of light in the first place.

An even more serious deficiency with Dyson's theory became apparent when it was discovered that the nose can easily differentiate molecules that have exactly the same chemical structure and identical Raman spectra, but are mirror images of each other. For example, the limonene molecule that is largely responsible for the smell of your orange can be described as a right-handed molecule. But there is a nearly identical molecule called dipentene which is its 'left-handed' mirror image molecule (see figure 5.3, where the pointed, shaded area at the bottom of each part represents the carbon–carbon bond that points either below (a) or above (b) the page). Dipentene has the same molecular bonds as limonene and thereby gives an identical Raman spectrum but its odour is very different: it smells like turpentine. Molecules that come in left- and right-handed forms are described as being *chiral*,* and they often have quite different

* A chiral molecule has a non-superposable mirror image.

155

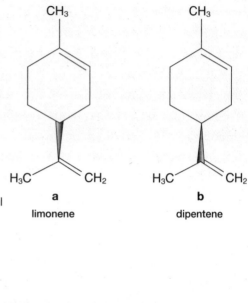

Figure 5.3: Limonene (a) and dipentine (b) are mirror-image molecules with very different smells. The molecules differ only in the orientation of the lower chemical group, which is represented as pointing below the page (bond pointing down) in limonene but above the page (bond pointing up) in dipentine. Of course, the dipentine molecule could be flipped over so that its chemical group points under the page like limonene, but then its double bond would flip round to the left, rather than the right, so it would still be different. The molecules are like left- and right-handed gloves.

odours. Another chiral compound is carvone, the chemical found in seeds such as dill and caraway, which is responsible for the caraway scent; its mirror-image molecule smells like spearmint. A Raman spectroscopist would be unable to distinguish these compounds with a spectrometer, but a simple sniff does it easily. Clearly, smell cannot rely, at least not solely, on detecting molecular vibrations.

These seemingly fatal flaws in the vibration theory of olfaction led to its eclipse by the lock and key theory through most of the latter half of the twentieth century, despite the best efforts of a few molecular vibration enthusiasts, such as the Canadian chemist Robert H. Wright, who provided a potential solution to the problem of left- and right-handed molecules that possess the same bonds but have different smells. He pointed out that the olfactory receptors are themselves likely to be chiral (coming in left- or right-handed forms), so they would hold an odorant molecule in a left-handed or right-handed way that would then present its bonds to the vibration detector differently. To take a musical analogy, the left-handed Jimi

Hendrix (representing the olfactory receptor) generally held his guitar (the chiral odorant molecule) with its neck pointing to the right; whereas the right-handed Eric Clapton held his guitar (representing the mirror-image molecule) with the neck pointing left.* Both musicians could play the same riff (generate the same vibrations) on mirror-image guitars; but the sound picked up by a fixed microphone (representing the vibration detector part of the olfactory receptor) placed, say, just to the left of each musician, would be subtly different because their strings (the molecular bonds) are in different locations relative to the microphone. Wright proposed that chiral olfactory receptors detect the vibration frequencies of chemical bonds, but only when the bonds are in the right position: he claimed that the receptors come in right- or left-handed forms, just like guitar players. But with still no idea of how the biological vibration detector would actually work, the vibration theory remained on the margins of olfactory science.

However, the shape theory also has its problems. As we have already discussed, it has difficulty explaining odorant molecules with very different shapes but the same odour, and vice versa. To tackle these problems, Gordon Shepherd and Kensaku Mori came up with what is sometimes called the 'weak shape', or odotope, theory in 1994.[5] The key difference between this and the classical shape theory lies in Shepherd and Mori's proposal that, rather than the shape of the entire molecule being recognized by olfactory receptors, the receptors need only identify the shape of the component chemical groups. For example, as we have already pointed out, all the musky-smelling compounds in figure 5.2 have an oxygen atom linked to a carbon atom by a double bond. The odotope theory proposes that it is the shape of these chemical substructures, rather than that of the entire molecule, that is recognized by olfactory receptors. This theory makes better sense of the analytical nature of scent, but it suffers

---

* In fact Hendrix generally played a right-handed guitar upside down but he reversed the strings so that high E would be in the same position as if he were playing a left-handed guitar.

Figure 5.4: Molecules with the same chemical basic parts – such as vanillin and isovanillin, shown here – can, nevertheless, have very different smells.

Figure 5.4: Molecules with the same chemical basic parts – such as vanillin and isovanillin, shown here – can, nevertheless, have very different smells.

Vanillin   Isovanillin

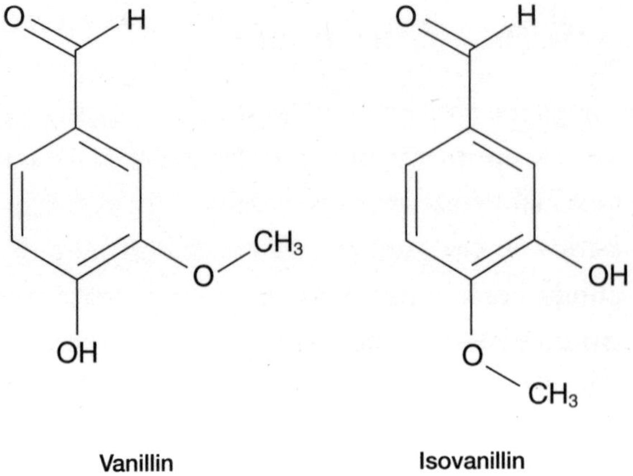

many of the same problems as the vibration theory when dealing with molecules containing the same chemical groups, but arranged differently. Thus, neither odotope nor vibration theory can explain how pairs of chemicals can have different odours despite possessing the same chemical groups arranged differently on the same molecular scaffold. For example, both vanillin (which forms the primary component of natural vanilla) and isovanillin consist of a six-carbon ring with three identical chemical groups attached at different positions (figure 5.4). The odotope theory would predict that the identical chemical groups should smell the same. Yet vanilla smells, well, like vanilla, but isovanillin has a nasty phenolic (sweet medicinal) scent.

To deal with these problems, shape theorists generally propose a combination of odotope theory and some sort of overall chiral shape recognition mechanism. Nevertheless, this still cannot explain the equally common situation of mirror-image molecules actually having the same smell.* It suggests that they are being recognized by the same receptor, which is the molecular equivalent of having the kind of hand that would fit both a left-handed and a right-handed glove. It just doesn't seem to make complete sense.

---

* For example, (4S,4aS,8aR)-(K)-geosmin and its mirror-image molecule (4R,4aR,8aS)-(C)-geosmin, which both smell 'earthy, musty'.

Smelling with a quantum nose

Shape recognition is intuitively easy to understand: we routinely deal with shape complementarity every time we slip on a glove, turn a key in a lock or use a spanner to tighten a nut. Enzymes (which we saw in action in chapter 3), antibodies, hormone receptors and other bio-molecules are also known to interact primarily via the geometric arrangement of their atoms and molecules; so it is not surprising that the shape theory of olfaction has received strong support from many biologists, including the olfaction receptor Nobel Prize winners, Richard Axel and Linda Buck.

Vibration-based communication is much less familiar to us, despite the fact that it is fundamental to at least two of our senses, vision and hearing. But whereas the physics of how the eye detects the vibration frequency of light and the ear records the vibration frequency of air are pretty well understood, no one had any idea, until recently, how the nose might detect the frequency of a molecular vibration.

Luca Turin was born in Lebanon in 1953 and studied physiology at University College London. After graduating, he moved to France to work at the National Centre for Scientific Research, and it was in Nice that he experienced an olfactory epiphany on a visit to the Galeries Lafayette. In the middle of the perfume room was a display mounted by the Japanese company Shiseido of their new perfume, Nombre Noir, which Turin describes: 'It was halfway between a rose and a violet, but without a trace of the sweetness of either, set instead against an austere, almost saintly background of cigar-box cedar notes. At the same time, it wasn't dry, and seemed to be glistening with a liquid freshness that made its deep colours glow like a stained-glass window.'[6] Turin's encounter with the Japanese scent was to inspire a lifelong quest to discover the secret of how molecules floating into the nose could create such evocative experiences.

Like Dyson before him, Turin was convinced that the correlations between vibrational spectra and scent couldn't be mere coincidence. He was persuaded by Dyson's argument that the olfactory receptors must somehow be detecting molecular vibrations. But, unlike Dyson, Turin proposed a speculative, yet plausible, molecular mechanism by which biomolecules could detect the vibrations of chemical bonds via quantum tunnelling of electrons.[7]

Tunnelling, you may remember from chapter 1, is the peculiar quantum mechanical property that arises from the ability of particles such as electrons or protons to behave as waves of probability capable of seeping through barriers that are impenetrable via any classical route. We discovered in chapter 3 how it plays a crucial role in many enzymes' reactions. While Turin was puzzling over the secret of scent, he came across a paper describing a new analytical chemical technique called *inelastic electron tunnelling spectroscopy* (IETS). In IETS, two metal plates are placed very close to one another, separated by a tiny gap. If a voltage is applied between the plates, electrons will gather on one plate, making it negatively charged (the donor) and will experience an attractive force from the other, positively charged, plate (the acceptor). Considered classically, the electrons lack the energy to jump across the insulating gap between the plates; but electrons are quantum objects and, if the gap is small enough, they can quantum tunnel across from donor to acceptor. This process is called elastic tunnelling because the electrons do not gain or lose energy in the process.

However, there is a crucial additional condition: an electron can tunnel elastically from its donor site across to the acceptor site only if there is an empty slot available for it at the exact same energy. If the nearest available gap in the acceptor is at lower energy, then the electron must lose some of its energy to make the jump. This process is called inelastic tunnelling. But the dumped energy needs to go somewhere, otherwise the electron can't tunnel. If a chemical is placed in the gap between the plates, then an electron can tunnel across so long as it is able to donate its excess energy to the chemical – which it can do so long as the molecules in the gap have bonds capable of vibrating at just the right frequency, corresponding to that of the dumped energy. Having passed on their excess energy in this way, these 'inelastically' tunnelling electrons arrive on the acceptor plate with slightly lower energy; so by analysing the energy differences between electrons leaving the donor site and arriving at the acceptor site, inelastic electron tunnelling spectroscopy probes the nature of a chemical's molecular bonds.

To return to our musical analogy, if you have ever played a stringed instrument you will know that it is possible to get a note out of a string without even touching it, by resonance. Indeed, this trick can be used to tune a guitar. If you fold a tiny scrap of very light paper over one of the strings, then pluck the same note on an adjacent string, you can make the paper scrap pop off without that string being touched at all. This is because, once you get the tuning just right, the plucked string sets the air vibrating, and the vibrating air passes the vibration on to the unplucked string, setting off its vibration in resonance with the plucked string. In IETS, the electron only pops off the donor site if the chemical between the two plates has a bond similarly tuned to just the right frequency for it to make the jump. In effect, the tunnelling electron loses energy by *plucking* a molecular bond on its quantum journey across the plates.

Turin proposed that olfactory receptors work in a similar way but with a single molecule – the olfactory receptor – taking the place of the IETS plates and gap. He envisaged an electron located first at a *donor site* in the receptor molecule. As in IETS, the electron could potentially tunnel to an acceptor site in the same molecule but, he proposed, it is prevented from doing so by an energy discrepancy between the two sites. However, if the receptor captures an odorant molecule that possesses a bond tuned to just the right vibrational frequency, then the electron can pop from donor to acceptor via tunnelling while simultaneously transferring just the right amount of energy to the odorant, effectively plucking one of its molecular bonds. Turin proposed that the tunnelled electron, now sitting in the acceptor site, causes the release of the tethered G protein molecular torpedo, causing the olfactory neuron to fire and thereby send a signal off towards the brain, allowing us to 'experience' the scent of the orange.

Turin managed to amass a lot of circumstantial evidence for his quantum vibrational theory. For example, as already mentioned, sulphur–hydrogen compounds usually have a strong rotten-egg smell, and they all possess a sulphur–hydrogen molecular bond that vibrates at around 76 terahertz (76 trillion oscillations per second). His theory makes a strong prediction: any other compound associated with a bond vibrational frequency of 76 terahertz should also have a rotten-egg smell, irrespective of its shape. Unfortunately, very few other compounds have that same vibration band in their spectra. Turin searched through the spectroscopy literature for a molecule with the same vibration. Finally, he discovered that the terminal boron–hydrogen bonds in chemicals called boranes have vibrations centred at 78 terahertz, which is quite close to the 76 terahertz S–H vibration. But what did boranes smell like? That information wasn't available in the spectroscopy literature, and the chemicals were so exotic that he couldn't get hold of any to have a sniff. But he found an old paper that described them as smelling repulsive, which is a term often used to describe a sulphurous smell. In fact, it turns out that boranes are the only known *non*-sulphur molecules that have the same rotten-egg stink as hydrogen sulphide: for example decaborane, which is made up of only boron and hydrogen atoms (chemical formula $B_{10}H_{14}$).

This discovery that among the literally thousands of chemicals that have been sniffed, the only one to stink like hydrogen sulphide is a molecule that shares the same vibrational frequency provided strong support for the vibrational theory of smell. Remember that perfumiers have been trying for decades to unlock the molecular key to scent. Turin had contrived to do what none of their chemists had managed: predict a scent from theory alone. It was the chemical equivalent of predicting the scent of a perfume from the shape of its bottle. Turin's theory also provided a biologically plausible quantum mechanism that would allow a biomolecule to detect a molecular vibration. But a 'plausible mechanism' is not enough. Was it correct?

Battle of the noses

The vibration theory had scored some encouraging successes, as with decaborane, but it still suffered from similar problems to the shape theory, such as the mirror-image molecules (e.g. limonene and dipentene) having very different scents, but identical vibrational spectra. Turin decided to test another prediction of his theory. You may remember that the tunnelling theory of enzyme action (chapter 3) was tested by replacing the most common form of hydrogen with one of its heavier isotopes, such as deuterium, in order to make use of the kinetic isotope effect. Turin tried a similar trick with an odorant called *acetophenone*, described as having a 'pungent sweet odour . . . resembling that of hawthorn or a harsh orange blossom'. He purchased a very expensive batch of the chemical in which each of the eight hydrogen atoms in its carbon–hydrogen bonds had been replaced with deuterium. The heavier atoms, like heavier guitar strings, will vibrate at lower frequencies: a normal carbon–hydrogen bond vibrates with a high note frequency of between 85 and 93 tera-hertz, but if deuterium is substituted for the hydrogen then the vibration frequency of the carbon–deuterium bond drops down to about 66 terahertz. The 'deuterated' chemical therefore has a very different vibrational spectrum from the hydrogen one. But does it smell any different? Turin locked the door of his laboratory before gingerly sniffing both compounds. He was convinced that they 'smelled different, the deuterated one less sweet, more solvent-like'.[8] Even after carefully purifying each compound he was convinced that the hydrogen and the deuterated forms smelled very different. His theory, he claimed, was vindicated.

Turin's research brought him to the attention of investors who provided the financial backing needed to set up a new company, Flexitral, devoted to exploiting his quantum vibration ideas to manufacture new fragrances. The author Chandler Burr even wrote a book describing Turin's quest for the molecular mechanisms of smell;[9] and the BBC filmed a documentary about his work.

But many were still far from convinced, particularly the shape theory enthusiasts. Leslie Vosshall and Andreas Keller from the Rockefeller University repeated the sniff tests with normal and deuterated acetophenone but, rather than relying on Turin's highly tuned nose, they asked twenty-four naïve subjects whether they could distinguish between the compounds. The results were unequivocal: no difference in smell. Their paper, published in *Nature Neuroscience* in 2004,[10] was accompanied by an editorial in which the vibration theory of scent was described as having 'no credence in scientific circles'.

But, as any medical researcher will tell you, trials in humans can be confounded by all sorts of complications, such as the expectations of the subjects and their experiences prior to the experiment. To avoid these problems, a team led by Efthimios Skoulakis of the Alexander Fleming Institute in Greece and including researchers from MIT, among them Luca Turin, decided to switch to a much better-behaved species: laboratory-bred fruit flies. The team devised the fruit-fly equivalent of Gabriele Gerlach's flume choice experiment with coral reef fish, described earlier in this chapter. They called it the fly 'T maze' experiment. The flies were introduced into a T-shaped maze through the stem and encouraged to fly to the junction where they would have to make a choice whether to go left or right. Scented air was pumped into each arm, so by counting the number of flies that went in each direction the researchers could work out whether the flies could distinguish between the odorants loaded respectively into the left and right airflows.

The group first investigated whether the flies could smell acetophenone. Indeed they could: just a dab of the chemical in the end of the right-hand arm of the maze was sufficient to persuade nearly all the flies to fly towards its fruity odour. The group then substituted deuterium for the hydrogen atoms in the acetophenone; but, in a new twist, they replaced either three, five or all eight hydrogen atoms with deuterium and tested each version of the chemical separately, with the undeuterated compound always in the other arm of the maze. Their results were remarkable. With only three deuterium atoms, the flies lost their preference for turning right at the junction and randomly went left or right. But when the researchers loaded the right arm with the five- or eight-atom-substituted chemical the flies resolutely turned left, away from the deuterated odorant. It seemed they could smell the difference between the normal and the heavily deuterated form of acetophenone, and now they didn't like what they smelled. The team tested two additional odorants and found that the flies could easily distinguish between hydrogen and deuterium forms of octanol, but not between the corresponding forms of benzaldehyde. To demonstrate that the flies were using their sense of smell to sniff out the deuterium bonds, the researchers also tested a mutant strain of fly that lacks functional olfactory receptors. As expected, these *anosmic** mutants were completely unable to distinguish between the hydrogen and deuterated odorants.

Using a Pavlovian conditioning setup, the researchers even managed to train the flies to associate certain forms of chemicals with punishment: a mild electric shock to their feet. The team were then able to perform an even more remarkable test of the vibration theory. They first trained flies to avoid compounds with the carbon–deuterium bond, with its characteristic vibration at 66 terahertz. They then wanted to discover whether this avoidance could be generalized to very different compounds that happened to possess a bond vibration at the same frequency. And it could. The team discovered that the flies trained to avoid compounds with the carbon–deuterium bond also avoided compounds called nitriles whose carbon–nitrogen bond vibrates at the same frequency, despite being chemically very different. The study provided strong support for a vibration component of olfaction, at least in flies, and was published in the prestigious science journal *Proceedings of the National Academy of Science* in 2011.[11]

---

* From *anosmia*, the inability to detect odours: a condition that, in humans, is usually associated with trauma to the nasal epithelium, though rare genetic forms are known.

The following year Skoulakis and Turin teamed up with University College London researchers to return to the delicate question of whether humans can also smell by vibrations. Rather than relying solely on Turin's highly sensitive sense of smell, the team recruited eleven sniffing subjects. They first confirmed the Vosshall and Keller result: their test subjects could not sniff out the carbon–deuterium bonds in acetophenone. But the team reckoned that with only eight carbon–hydrogen bonds, the signal from the deuterated form of the chemical might be rather weak and thus indistinguishable to the average nose; so they decided to investigate more complex musky-smelling molecules (like those in figure 5.2) that have up to twenty-eight hydrogen atoms, all of which could be replaced with deuterium. This time, in contrast to the acetophenone trial, all of their eleven subjects could easily distinguish between normal and fully deuterated musk. Maybe humans really can sniff out molecules tuned to different frequencies after all.

Physicists take a sniff

One of the criticisms levelled at the quantum vibration theory was that its theoretical foundation was all rather vague. This has now been addressed by a team of physicists from University College London who, in 2007, carried out the 'hard-nosed' (if you'll pardon the pun) quantum calculations behind the tunnelling theory and concluded that it was 'consistent both with the underlying physics and with observed features of smell, provided the receptor has certain general properties'.[12] One of the team, Jenny Brookes, even went on to propose a solution to that niggling problem of mirror-image molecules such as limonene and dipentene (figure 5.3) that have the same vibrations but very different odours.

In fact, it was Jenny's supervisor and mentor, the late Professor Marshall Stoneham, who first came up with what is sometimes called the swipe-card model. Stoneham was one of the leading UK physicists of his generation with interests that ranged from nuclear safety to quantum computing, biology and, appropriately for this chapter, music: he played the French horn. Their theory is a quantum mechanical elaboration on Robert H. Wright's idea that both the shape of the olfactory receptor and the bond vibrations of the odorant molecule play a role in smell. They proposed that the binding pocket of the olfactory receptor works like a swipe-card machine. Swipe cards have a magnetic strip that is read to generate an electric current in the swipe-card machine. But not everything fits into a swipe-card reader: the card has to be the right shape and thickness, with its magnetic strip in the right place, before you can even *use* it and check whether the machine recognizes it. Brookes and her colleagues proposed that olfactory receptors work in a similar way. An odorant molecule, the team postulated, must first fit into a left- or right-handed chiral binding pocket, rather like a credit card fitting in a card reader. So odorants with the same bonds but different shapes, such as a left-handed and a right-handed version of the same molecule, will be picked up by different receptors. Only after either odorant has fitted into its complementary receptor does it have the potential to stimulate the vibration-induced electron tunnelling event to make the receptor neuron fire; but because the left-handed molecule will be firing a left-handed receptor, it will smell different from a right-handed molecule firing a right-handed receptor.

If we return to our musical analogy one final time, with the guitar acting as the odorant molecule and the guitar strings as the molecular bonds that need to be plucked, then the receptors come in Eric Clapton or Jimi Hendrix forms. Both can play the same molecular notes, but right- or left-handed molecules have to be picked up by right- or left-handed receptors, just as right-handed guitars have to be played by right-handed guitarists. So although limonene and dipentene have the same vibrations, they have to be held by left- or right-handed olfactory receptors. The different receptors will be wired to different regions of the brain and will thereby generate different smells. This combination of shape and quantum vibration recognition at last provides a model that fits nearly all the experimental data.

Of course, the fact that this model fits the data doesn't itself prove that there is a quantum basis for olfaction. The experimental data provide strong evidence for any theory of olfaction that involves both shape and vibration. No experiment has yet directly tested whether quantum tunnelling is involved in smell. However, so far at least, inelastic quantum tunnelling by electrons is the only known mechanism that provides a plausible explanation for how proteins can detect vibrations in odour molecules.

The vital piece of the olfaction puzzle that is still missing is the structure of olfactory receptors. Knowing this would make it easier to find the answer to key questions such as how snugly the binding pockets are tailored to each odorant molecule, whether mirror-image molecules bind to the same receptors, and whether the receptor molecules possess electron donor and acceptor sites suitably positioned to promote inelastic electron tunnelling. Yet despite many years of effort by some of the top structural biology groups around the world, no one has yet managed to isolate olfactory receptor molecules susceptible to study in the same way that allowed quantum mechanical mechanisms to be elucidated in enzymes (chapter 3) or photosynthetic pigment proteins (chapter 4). The problem is that in its natural state the olfactory receptor is embedded in the cell's membrane, a bit like a jellyfish floating on the surface of the sea. Taking the receptor protein out of the membrane is like taking a jellyfish out of the ocean: it is just not going to retain its shape. And no one has yet found a way of determining the structure of proteins while they remain embedded in cell membranes.

So, although considerable controversy remains, the only theory that provides an explanation of how flies and humans can distinguish the smells of normal and deuterated compounds is based on the quantum mechanical mechanism of inelastic electron tunnelling. Experiments have recently shown that, as well as flies and humans, other insects and even fish are able to sniff the differences between hydrogen and deuterium bonds. If quantum smelling is found in such a range of creatures it is likely to be very widespread. Humans, fruit flies, anemonefish and a host of other animals are probably harnessing the ability of an electron to vanish from one point in space and instantly materialize in another so that they can capture that 'message from a material reality' and find food, or a mate – or their way home.

6

The butterfly, the fruit fly
and the quantum robin

$\mathbf{B}$ ORN IN Toronto, Canada, in 1912, Fred Urquhart went to a school that bordered a cat-tail marsh. There he spent endless hours observing the insects, particularly the butterflies that populated the reed beds. His favourite time of the year was the early summer, when the marsh saw the arrivals of thousands of monarchs, those iconic North American butterflies with their familiar orange-and-black wing pattern. Monarchs would remain for the summer, feeding on the native milkweeds, before flying off again in the autumn. And the question that particularly intrigued Fred was: where do the butterflies go?

As St Paul is reputed to have said, adults generally put away their childish things. Not Fred, however, who continued, as he grew up, to wonder where the monarchs spent the winter. After studying zoology at the University of Toronto and eventually becoming a professor in the subject, he returned to his childhood question. By this time he had married Norah Patterson, a fellow zoologist and butterfly-lover.

Using the classic animal tagging techniques, Fred and Norah now attempted to discover the secret of the monarchs' disappearance. This wasn't easy. While tags tied to the feet of robins, or pinned to the fins of whales, work fine, attaching a tag to the delicate membranous wings of a butterfly presents a wholly different challenge. The husband-and-wife team experimented with sticky labels and gluing tags onto the insects' wings; but either the tags fell off or the tagged butterflies had trouble flying. It wasn't until 1940 that they hit upon the solution: a tiny adhesive label similar to those that are so difficult to scrape off newly bought glassware. Armed with this device, they began tagging and releasing hundreds of monarch butterflies, each with an identifying number and the instructions that, if it were found, the finder should 'Send to Zoology, University of Toronto'.

But there were millions of monarchs in America and only two butterfly-loving Urquharts. So the couple took to recruiting volunteers, and by the 1950s had marshalled a network of thousands of butterfly enthusiasts who in turn had tagged, released, captured and recorded hundreds of thousands of butterflies. As Fred and Norah constantly updated a map that tracked these capture and release locations, a pattern gradually emerged. The butterflies setting off from the Toronto area tended to be captured along a diagonal flight path travelling southwards that crossed the United States from the north-east to the south-west, passing through Texas. But despite numerous field trips, the Urquharts could not identify the final destination of these wintering butterflies in the southern US states.

Eventually, the Urquharts turned their eyes further south, and in 1972 a frustrated Norah wrote about their project to newspapers in Mexico, asking for volunteers to report any sightings and to help with the tagging. In February 1973 a letter arrived from a Kenneth C. Brugger of Mexico City offering to help. With his dog, Kola, Ken took up the quest, driving his camper van into the Mexican countryside in the late evenings in search of butterflies. Over a year later, in April 1974, he reported having seen large numbers of monarchs in the Sierra Madre mountain range of central Mexico. Then, late that year, Ken reported spotting the bodies of many tattered and dead butterflies along roads in the Sierras. Norah and Fred wrote back that they believed that flocks of birds must have been feeding on large swarms of passing monarchs.

On the evening of 9 January 1975, Ken telephoned the Urquharts in some excitement with the news that he had 'found the colony! . . . millions of monarchs – in evergreens beside a mountain clearing'.

Ken told them that he had received a tip-off from Mexican woodcut-ters who claimed to have seen swarms of red butterflies as they trailed across the mountain with their laden donkeys. With the support of the National Geographical Society, Norah and Fred then put together an expedition to find and record the elusive wintering home of the monarchs, arriving in Mexico in January 1976. The following day they drove out to a village from which they set off to hike up the 'Mountain of the Butterflies', an ascent of 10,000 feet. Such an ardu-ous climb at altitude was not a trivial undertaking for a couple now in their senior years (Fred was sixty-four), and they were quite con-cerned about whether they could make it to the top. Nevertheless, with pounding hearts and memories of brightly coloured butterflies fluttering in the Toronto sunshine, they reached the summit, a plat-eau sparsely forested with juniper and holly. There were no butterflies. Disappointed and exhausted, they climbed down into a clearing filled with oyamel, a type of fir tree native to the mountains of cen-tral Mexico – and it was here that Fred and Norah finally found what they had sought for half a lifetime: 'Masses of butterflies – everywhere. In the quietness of semi-dormancy, they festooned the tree branches, they enveloped the oyamel trunks, they carpeted the ground in their tremendous legions.' As they stood gaping at the incredible spectacle, a tree branch broke off and there, among the debris of dislodged butterflies Fred spotted the familiar white tag with its instruction: 'Send to Zoology, University of Toronto'. That particular butterfly had been tagged by a volunteer named Jim Gil-bert in Chaska, Minnesota, more than two thousand miles away![1]

The voyage of the monarch butterfly is now recognized as one of the great animal migrations of the world. Between September and November every year, millions of monarchs in south-east Canada head south-west on a journey of several thousand miles that will take them across desert, prairie, fields and mountains, on the way thread-ing through the geographical needle's eye of a 50-mile-wide gap of cool river valleys between Eagle Pass and Del Rio in Texas, eventually to roost on the peaks of only a dozen or so high mountains in central

Mexico. And then, after overwintering on the cool Mexican mountaintops, the monarchs undertake the reverse journey in the spring to return to their summer feeding grounds. Most remarkably, no single butterfly makes the entire journey. Instead, they breed en route so that the butterflies that return to Toronto are the grandchildren of the monarchs that first left Canada.

How do these insects navigate with such accuracy that they can reach a tiny target thousands of miles from their origin, one that only their ancestors had previously visited? This is another of those huge mysteries of nature that is only now beginning to be unravelled. Like all migrating animals, the butterfly uses a variety of senses involving sight and smell, including a sun compass that can correct for the moving position of the sun during the day via its *circadian clock*, the biochemical process in all animals and plants that oscillates within a 24-hour period, tracking the day–night cycle.

Circadian clocks are familiar to us as the source of our feeling tired at night and wakeful in the morning; and of our suffering jet lag when their rhythms are disturbed by long-distance air travel. The last couple of decades or so has seen a succession of fascinating discoveries about how they work. One of the most surprising is the finding that subjects who are kept in isolation in constant light conditions still manage to maintain a roughly 24-hour cycle of activity and rest despite having no external cues. It seems that our body clock, our circadian clock, is hardwired. This built-in clock, the body's 'pacemaker' or circadian sense, is located in the hypothalamus gland buried inside the brain. But although subjects kept in constant light conditions still maintain a roughly 24-hour cycle, their circadian clock gradually drifts away from the actual times of the day, so their periods of wakefulness and sleep will not be in synch with those of people outside the study. Yet once exposed to natural light, the subject's body clock soon recalibrates to the actual light–dark cycle in a process known as *entrainment*.

The monarch butterfly's sun compass works by comparing the height of the sun with the time of day – a relationship that varies with both latitude and longitude. It must also have a body clock that, like our own, is similarly automatically entrained by light, to compensate for the changing times of sunrise and sunset during its long migration. But where does the monarch house its circadian sense?

As the Urquharts discovered, butterflies are not the easiest animals to work with; the fruit fly, *Drosophila*, which we encountered in the last chapter sniffing its way through a maze, is a much more convenient laboratory insect as it breeds very rapidly and can easily mutate. Like us, fruit flies adjust their circadian rhythms to the cycles of light and dark. In 1998, geneticists found a fruit-fly mutant whose circadian rhythm could not be affected by exposure to light.[2] They discovered that the mutation was in a gene encoding an eye protein called cryptochrome. Rather like protein scaffolds in the photosynthetic complexes that hold chlorophyll molecules together (as we saw in chapter 4), the cryptochrome protein is wrapped around a pigment molecule called FAD (*flavin adenine dinucleotide*) that absorbs blue light. Just like in photosynthesis, the light absorption knocks an electron out of the pigment, which leads to the generation of a signal that travels to the fly's brain to keep its body's clock in synch with the daily light–dark cycle. The mutant flies discovered in 1998 had lost this protein, so their body clocks were no longer adjusting to the cyclical change between light and dark: they had lost their circadian sense.

Similar cryptochrome pigments were later found in the eyes of many other animals, including humans, and even in plants and photosynthetic microbes, where they help to predict the time of day best suited for photosynthesis. They may represent a very ancient light-detection sense that evolved in microbes billions of years ago to synchronize cell activities with diurnal rhythms.

Cryptochrome is also found in monarch butterflies' antennae. This was initially puzzling: what was an eye pigment doing in antennae? But insect antennae are truly amazing organs that house multiple senses, including those for smell and hearing, the detection of air pressure and even of gravity. Could they also house the insect's circadian sense? To test this hypothesis, scientists painted some butterfly antennae black, thereby preventing them from receiving light signals. What they discovered was that the butterflies with blackened antennae could no longer entrain their sun compass with the cycle of night and day: they had lost their circadian sense. So the butterfly's antennae seemed to house its biological clock. Remarkably, the clock in the butterfly's antennae could be entrained by light even when removed from the rest of the insect's body.

Was cryptochrome responsible for the monarch's light entrainment? Unfortunately, it isn't as easy to mutate butterfly genes as it is those of the fruit fly, so in 2008 Steven Reppert and colleagues from the University of Massachusetts did the next best thing. The team replaced the defective cryptochrome gene in mutant fruit flies with the monarch butterfly's *healthy* gene and showed that it restored the fly's ability to entrain its circadian rhythms with light.[3] If the butterfly cryptochrome managed to keep fruit flies on time, then it was very likely to be doing the job of setting the monarch's all-important body clock so it could fly all the way from Toronto to Mexico without getting lost.

But what has any of this to do with quantum mechanics? The answer has to do with another aspect of animal migration, namely the sense we call 'magnetoreception' – the ability to detect the earth's magnetic field. As we saw in chapter 1, it has been known for a while that many creatures, including fruit flies and butterflies, possess this capability, and magnetoreception, particularly in robins, has become the poster child of quantum biology. By 2008 it was clear that the robin's magnetic sense involved light (more on this later), but the nature of the light receptor was elusive. Steven Reppert wondered whether the cryptochrome that provided flies with the light sensitivity that helped to entrain their circadian rhythms could also be involved in their magnetoreception sense. To test the theory he performed the kind of flume choice experiment that Gabriele Gerlach had used to demonstrate olfactory navigation in clownfish (chapter 5), in which the test animal is forced to use sensory cues to choose between two routes to its food.

The researchers found that the flies could be trained to associate a sugar reward with the presence of a magnetic field. When given the option to fly down either the magnetized or the non-magnetized arm of a maze (without food, so without olfactory cues), they chose the magnetized path. The flies must sense the magnetic field. So was cryptochrome involved? The researchers found that mutant fruit flies genetically engineered to lack cryptochrome were equally likely to fly down either arm of the maze, demonstrating that cryptochrome was essential for their magnetic sense.

In their 2010 paper, Reppert's group also demonstrated that the flies kept their magnetic sense when their cryptochrome gene was replaced by the gene encoding cryptochrome from monarch butterflies[4] – showing that the monarch butterfly may well also use cryptochrome to detect the earth's magnetic field. In fact, a paper from the same group in 2014 demonstrated that, like the European robin we met in chapter 1, the monarch butterfly possesses a light-dependent inclination compass which it uses to find its way from the Great Lakes to a Mexican mountaintop; and, as expected, it appears to be housed in its antennae.[5]

But how does a light pigment also detect an invisible magnetic field? To answer that question we have to return to our friend the European robin.

The avian compass

As we pointed out in chapter 1, our planet is a giant magnet, with a magnetic field of influence that extends from its inner core all the way out into space for thousands of miles. This magnetized bubble, the 'magnetosphere', protects all life on earth, because without it the solar wind – the stream of energetic particles emitted from the sun – would have long ago eroded our atmosphere. And, unlike the magnetism of a typical bar magnet, the earth's field changes over time, because it has its origins inside the earth's molten iron core.

The precise origin of this magnetism is complicated, but it is thought to be due to what is known as a geo-dynamo effect, whereby electric currents are generated by the circulation of liquid metals in the earth's core, which in turn generates a magnetic field.

So, life on earth owes its existence to this protective magnetic shield. But its usefulness to living creatures doesn't end there; scientists have known for over a century that many species have evolved ingenious ways of making use of it. Just as human sailors have used the earth's magnetic field for thousands of years to navigate the oceans, so many of earth's other creatures, including marine and terrestrial mammals, birds (such as our robin) and insects, have evolved over millions of years a sense that detects the earth's magnetic field and uses it to navigate.

The earliest evidence of this capability was provided by a Russian zoologist, Aleksandr von Middendorf (1815–94), who recorded the places and dates of arrival of several species of migratory birds. On the basis of these data he drew a number of curves on a map, which he referred to as *isepipteses* (lines of simultaneous arrival). From these, which reflected the directions of arrival of the birds, he deduced that there was 'a general convergence northwards' towards the magnetic north pole. When he published his findings in the 1850s he proposed that migratory birds orientate themselves by the earth's magnetic field, referring to them as 'sailors of the air' that can navigate 'in spite of wind, weather, night or cloud'.[6]

Most other nineteenth-century zoologists remained sceptical. Paradoxically, even those scientists who were prepared to accept more outlandish pseudoscientific notions like paranormal activity – and there were many prominent scientific names in the late nineteenth century who did so – could not believe that magnetic fields could influence life. Joseph Jastrow, for example, an American psychologist and psychic researcher, in July 1886 published a letter in the journal *Science* entitled 'The existence of a magnetic sense'. He described experiments he had carried out to test whether humans could be in any way affected by a magnetic field, but had to report that he found no sensitivity whatsoever.

Yet if you fast-forward from Jastrow into the twentieth century you encounter the work of Henry Yeagley, an American physicist who carried out research for the US Army Signal Corps during the Second World War. Avian navigation was of interest to the military because homing pigeons were still being used to carry messages and aviation engineers hoped to learn from their navigational capabilities. Yet how the birds managed to find their way home so unerringly remained a mystery. Yeagley developed a theory that homing pigeons could sense both the earth's rotation and its magnetic field. This, he claimed, would create a 'navigational grid work' in the bird's brain, giving it both longitude and latitude coordinates. He even tested his theory by attaching small magnets to the wings of ten pigeons and non-magnetic strips of copper of the same weight to ten others. Eight of the ten birds with copper strips attached to their wings found their way home, but only one of the ten pigeons with magnets attached to their wings managed to reach their nest. Yeagley concluded that the birds utilize a magnetic navigational sense to navigate, which could be disrupted by magnetic strips.[7]

Although Yeagley's experimental results were initially dismissed as far-fetched, several researchers have since established beyond reasonable doubt that a wide range of animals have an inbuilt sensitivity to the earth's magnetic field, giving them an acute sense of direction. Sea turtles, for example, are able to return to the same breeding beach, thousands of kilometres away from their ocean feeding grounds, without any visual landmarks; and researchers have shown that their navigational sense is impaired if powerful magnets are attached to their heads. In 1997, a team at the University of Auckland in New Zealand published research in *Nature* suggesting that the rainbow trout uses magnetoreceptor cells located in its nose.[8] If proved correct, this would be the first example of a species that is able to *smell* the direction of the earth's magnetic field! Microbes use the earth's magnetic field to help navigate through murky water; and even organisms that don't migrate, such as plants, appear to retain a magnetoreception sense.

The ability of animals to detect the earth's magnetic field is no longer in doubt. The mystery is how they do it, not least because the earth's magnetic field is extraordinarily weak and would not normally be expected to influence any chemical reactions in the body. There are two principal theories, and both are likely to be involved in different animal species. The first is that the sense functions like a conventional magnetic compass, while the second is that magneto-reception is conferred by a chemical compass.

This first idea, that a form of conventional compass mechanism resides somewhere in an animal's body, was bolstered by the discovery of tiny crystals of magnetite, the naturally occurring magnetic iron oxide mineral, in many of the animals and microbes that seemed to possess a magnetic sense. For example, the bacteria that utilize a magnetic sense to orientate themselves in muddy marine sediments are often filled with bullet-shaped crystals of magnetite.

By the late 1970s, magnetite had been detected in the bodies of various animal species known to navigate with the help of the earth's magnetic field. Notably, it seemed to have been found inside neurons within the upper beaks of the most famous of avian navigators, homing pigeons,[9] suggesting that their neurons were responding to magnetic signals picked up by the magnetite crystals and then sending a signal to the animal's brain. More recent research showed that pigeons became disorientated and lost their ability to track the geomagnetic field when small magnets were attached to their upper beaks, where those magnetite-filled neurons were apparently located.[10] It seemed that the origin of a magnetoreceptive sense had finally been located.

However, it was back to the drawing boards in 2012 when yet another paper appeared in *Nature* describing a detailed 3D study of the pigeon's beak using an MRI scanner, which concluded that those magnetite-containing cells in the pigeon's beak almost certainly had nothing to do with magnetoreception at all but were in fact iron-rich cells called macrophages that are involved in immunity to pathogens, not, as far as is known, sensory perception.[11]

It's at this point that we should rewind the clock and return to that remarkable German ornithologist, Wolfgang Wiltschko, whom we first met in chapter 1. Wiltschko's interest in bird navigation began in 1958 when he joined a Frankfurt-based research group run by the ornithologist Fritz Merkel. Merkel was one of the few scientists at the time studying the magnetic sense of animals. One of his students, Hans Fromme, had already shown that some birds could orientate themselves inside featureless closed rooms, which demonstrated that their navigational capability was not based on visual clues. Fromme had proposed two possible mechanisms: either the birds were receiving some sort of radio signals from the stars or they could sense the earth's magnetic field. Wolfgang Wiltschko suspected the latter.

In the autumn of 1963, Wiltschko began conducting experiments with European robins, which as you may remember normally migrate between northern Europe and North Africa. He placed robins, captured in mid-migration, inside magnetically shielded chambers and then exposed the birds to a weak, artificial, static magnetic field generated by a device called a Helmholtz coil that can mimic the earth's geomagnetic field but whose strength and orientation can be changed. What he found was that those birds captured during migration in the autumn or the spring became restless and would cluster on the side of the chamber that coincided with their migratory direction relative to the artificial field. After two years of painstaking effort he published findings in 1965 demonstrating that the birds were sensitive to the direction of the applied field and so, he surmised, could similarly detect the earth's magnetic field.

These experiments conferred a degree of respectability on the idea of avian magnetoreception and sparked further research. But, at the time, no one had the faintest idea how this sense worked – how the extremely weak magnetic field of the earth could actually influence the bodies of animals. Scientists couldn't even agree where in an animal's body the magnetoreception sensory organ was situated.

Even after magnetite crystals were found in several animal species, implying a conventional magnetic compass mechanism, the robin's navigational capability remained a mystery because no magnetite could be detected in the bird's body. The robin's sense also displayed several puzzling features that didn't fit with a magnetic compass, not least because the birds lost their ability when they were blindfolded, indicating that they need to 'see' the earth's magnetic field. But how does any animal *see* a magnetic field?

It was in 1972 that the Wiltschkos (Wolfgang having by this time teamed up with his wife Roswitha) discovered that the robin's compass was unlike any that had been previously studied. A normal compass has a magnetized needle, one end of which (its south pole) is attracted towards the magnetic north pole of the earth, while the other end points towards the south pole. But there is a different kind of compass that doesn't discriminate between the magnetic poles. This, you may remember from chapter 1, is called an *inclination compass*; and it points to whichever pole is nearest, so it can only tell you if you are heading either towards or away from that pole, whichever one it is. One way of providing this kind of information is to measure the angle of the earth's magnetic field lines with respect to the surface of the earth (figure 6.1). This angle of inclination (hence the name

Figure 6.1: The earth's magnetic field lines and the angle of inclination.

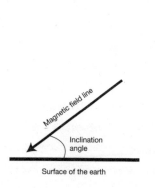

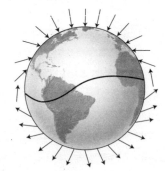

for this kind of compass) is near-vertical (pointing into the ground) close to the poles, but parallel to the ground at the equator. Between the equator and the poles the magnetic field lines enter the earth at some angle less than 90° and that angle points towards the nearest pole. So any device that measures this angle can function as an inclination compass and provide directional information.

In their 1972 experiments the Wiltschkos trapped the test birds in a shielded chamber and subjected them to an artificial magnetic field. Crucially, reversing the polarity of the field, by turning the magnet round by 180°, had no effect on their behaviour: the birds would orientate themselves in relation to the closest magnetic pole, whichever one it happened to be; so they didn't possess a conventional magnetic compass. That 1972 paper established that the robins' magnetoreceptor was indeed an inclination compass. But how it worked remained a mystery.

Then in 1974 Wolfgang and Roswitha were invited to Cornell University in the United States by the American bird migration expert Steve Emlen. In the 1960s he had developed with his father John, also a highly respected ornithologist, a special bird chamber that became known as an Emlen funnel.* Shaped like an inverted cone, this funnel has an inkpad at the bottom and blotting paper on the interior sloping sides (figure 6.2). When a bird hops or flutters up the sloping walls it leaves telltale footprints that give information about the preferred direction in which it would fly if it could escape. The bird species the Wiltschkos studied at Cornell University was the indigo bunting, a small North American songbird that, like the European robin, migrates using some kind of internal compass. Their year-long study of this bird's behaviour inside the Emlen funnel was published in 1976,[12] and established beyond doubt that the indigo bunting, like the robin, was able to detect the geomagnetic field. Wolfgang Wiltschko regards the publication of this first Cornell-based paper as the team's breakthrough moment, for it established

---

* Not to be confused with Emlen Tunnell, the great American football player of the 1950s.

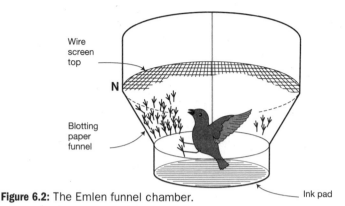

Figure 6.2: The Emlen funnel chamber.

beyond doubt that migratory birds have a built-in magnetic compass and caught the attention of many of the world's leading ornithologists.

Of course, no one in the mid-1970s had a clue how a biological magnetic compass might work. However, as we saw in chapter 1, in the same year that the Wiltschkos and Stephen Emlen published their work, the German chemist Klaus Schulten proposed a chemical mechanism that links light with magnetoreception. Schulten had recently graduated from Harvard with a PhD in chemical physics and returned to Europe, where he obtained a position at the Max Planck Institute for Biophysical Chemistry in Göttingen. There he became interested in the possibility that electrons generated in the fast triplet reaction by exposure to light could be quantum entangled. His calculations suggested that if entanglement was indeed involved in chemical reactions then the speed of these reactions should be affected by an external magnetic field, and he proposed a way of proving his theory.

As he talked freely about his idea, Schulten developed a reputation at the Max Planck Institute for being regarded as somewhat crazy. His problem was that he was a theoretical physicist who worked with paper, pen and computers, not a chemist; and certainly not an experimental chemist capable of donning a lab coat and performing the kind of experiment that would prove his ideas. Thus he was in the position of many theoreticians who come up with a neat idea but have then to find a friendly experimentalist willing to take time out of their busy lab schedule to test a theory that, more often than not, will prove to be wrong. Schulten had no luck in persuading any of his chemist colleagues to try out his idea, because none of them believed that his proposed experiment had any chance of success.

The source of all this scepticism, Schulten discovered, was the institute's lab manager, Hubert Staerk. Eventually Schulten summoned up the courage to confront Staerk in his office, where he finally learned the reason for this entrenched scepticism: Staerk had already done the experiment and found no effect of magnetic fields. Schulten was thunderstruck. It seemed that his hypothesis was to suffer the fate described by the evolutionary biologist Thomas Huxley as just another 'beautiful theory . . . killed by an ugly fact'.

After thanking Staerk for conducting the experiment, the dejected Schulten was about to leave his office, but then turned back and asked to see the disappointing data. When Staerk showed him the file, Schulten's mood suddenly lifted. He noticed something that Staerk had missed: a small but significant blip in the data that he had perfectly predicted. He recalls that it was 'exactly what I expected, and so I was very happy that I saw it. A disaster turned into a happy moment, because I knew what to look for. He didn't.'[13]

Schulten immediately set to writing what he was sure would be a breakthrough scientific paper – but he was soon to get another shock. Sharing a drink at a conference with a colleague, Maria-Elisabeth Michel-Beyerle, from the Technical University of Munich, he discovered that Michel-Beyerle had done the exact same experiment. This put Schulten in an ethical quandary. He could reveal his discovery and potentially prompt Michel-Beyerle to rush back to Munich to write her own paper, which might scoop his own publication; or he could make his excuses and high-tail it back to Göttingen to write up his own results. But if he did flee without saying a word and then published first, Michel-Beyerle might later accuse him of stealing her idea. He recalls his thoughts: 'If I don't now tell what I know, she may say I went home to do the experiment.'[14] In the end, Schulten came clean and admitted to Michel-Beyerle that he had done similar work. Both scientists stayed for the remainder of the conference and then returned to their respective homes to write their own papers (Schulten's appeared just a little before Michel-Beyerle's) describing the discovery that the weird property of quantum entanglement can indeed influence chemical reactions.

Schulten's 1976 paper[15] proposed that quantum entanglement was responsible for the speed of the exotic fast triplet reactions studied in the Max Planck laboratory; but his groundbreaking paper also presented Staerk's experimental data, which clearly showed that the chemical reaction was sensitive to magnetic fields. With two big results 'in the bag', many scientists would have been content; but Schulten, not yet thirty, still possessed the recklessness of youth and was prepared to stick his neck out yet further. Aware of the Wiltschkos' robin migration work and the problem of finding a plausible chemical mechanism for a biological compass, he realized that his spinning electrons could provide such a mechanism; and in a 1978 paper he proposed that the avian compass depended on a quantum-entangled radical pair mechanism.

At the time, hardly anyone took this idea seriously. Schulten's colleagues at the Max Planck Institute considered it to be just another of his crazy notions, and the editors of *Science*, the top scientific journal to which he first sent his paper, were similarly unimpressed, writing: 'A less bold scientist may have designated this idea to the waste paper basket.'[16] Schulten describes his response: 'I scratched my head and thought, "This is either a great idea or entirely stupid." I decided it was a great idea and published it quickly in a German journal!'[17] But at this juncture most scientists, if they knew about it at all, filed Schulten's speculative theory away with pseudoscientific and paranormal explanations of magnetoreception.

Before we can see how Schulten's and the Wiltschkos' work might help to explain how birds find their way around the globe, we need to return to the mysterious quantum world and take a careful look at the phenomenon of entanglement, which we described briefly in the first chapter of this book. You may remember that entanglement is so strange that even Einstein insisted that it could not be correct. First, however, we need to introduce you to another peculiar property of the quantum world: 'spin'.

Quantum spin and spooky action

Many popular science books on quantum mechanics use the concept of 'quantum spin' to highlight the strangeness of the subatomic world. We've chosen not to do so here simply because it is probably the notion furthest removed from anything that we can conceptualize using everyday language. But we cannot put the task off any longer, so here goes.

Just as the earth spins on its axis as it orbits the sun, so electrons and other subatomic particles have a property called *spin* that is distinct from their normal motion. But, as we hinted in chapter 1, this 'quantum spin' is unlike anything that we can visualize on the basis of our everyday experience of spinning objects like tennis balls or planets. For a start, it doesn't really make sense to talk about the speed of an electron spinning, as its spin can only take on one of two possible values: it is quantized, just as energy is quantized at the quantum level. Electrons can only – in a loose sense – spin in either a clockwise or an anticlockwise direction, corresponding to what is usually referred to as spin 'up' or spin 'down' states. And because this is the quantum world, an electron can, when not being watched, *spin in both directions at the same time*. We say that their spin state is a superposition (i.e. combination or mixture) of spin-up and spin-down. In a sense, this may sound even weirder than saying that an electron can be in two locations at once – for how can a single electron spin both clockwise and anticlockwise at the same time?

And just to stress how counterintuitive this notion of quantum spin is, what we regard as a 360° rotation will not take an electron back to its original state; to do that, it needs to make two full rotations. This sounds strange because we still tend to think of an electron as a tiny sphere, maybe something like a very small tennis ball. But tennis balls are inhabitants of the macroscopic world, and electrons live in the subatomic quantum world where the rules are different. In fact, electrons are not only *not* tiny spheres, they cannot even be said to have a size at all. So, while quantum spin is just as 'real' as the rotation of a tennis ball, it doesn't have a counterpart in the familiar everyday world and cannot be pictured.

However, do not therefore think that this is just an abstract mathematical concept that exists only in textbooks and impenetrable physics lectures. Every electron in your body, and everywhere else in the universe, spins in this peculiar way. In fact, if they didn't, the world as we know it, including us, just couldn't exist, because quantum spin plays a key role in one of the most important ideas in science, namely the Pauli Exclusion Principle, which underpins the whole of chemistry.

One of the consequences of the Pauli Exclusion Principle is that if two electrons are paired up in an atom or molecule and have the same energy (remember from chapter 3 that the chemical bonds that hold molecules together are made up of electrons that are shared between atoms), then they have to have opposite spin. We can then think of their spins as cancelling out, and we refer to them as being in a *spin singlet state*, since they can only inhabit a single state. This is the normal state of pairs of electrons in atoms and most molecules. However, when not paired together at the same energy level, two electrons can spin in the same direction, and this is called a *spin triplet state*,* as in the reaction that Schulten studied.†

You may be familiar with highly dubious claims that identical

---

* The term 'triplet' here can be confusing to the non-expert in quantum mechanics, especially since it refers to just a pair of electrons, so here is a very brief explanation: an electron is said to have a spin of ½. So, when a pair of electrons have opposite spin, these values cancel: ½ − ½ = 0. This is referred to as a spin singlet state. But when they have their spins pointing in the same direction, these values add up: ½ + ½ = 1. The term 'triplet' refers to the fact that a combined spin of 1 can be pointing in three possible directions (up, down and sideways).
† The two unpaired electrons in an oxygen molecule that hold its two atoms together are normally in a spin triplet state.

twins are able to sense each other's emotional states even when separated by vast distances. Somehow, the idea goes, twins are joined at a psychic level that science has yet to understand. Similar claims have been made to explain how a dog apparently senses when its owner is coming home. We should clarify that neither of these examples has any scientific merit, even though some people have mistakenly tried to ascribe to them a quantum mechanical basis. However, although such 'instantaneous action at a distance' (as it is often described) is not found in our everyday classical world, it is a key feature of the quantum domain. Its technical name is *nonlocality*, or *entanglement*, and it refers to the idea that something happening 'over here' can have an *instantaneous* effect 'over there' no matter how far away 'there' is.

Consider a pair of dice. The mathematical probability of throwing a double is easy to work out. For any given number that one of the dice lands on, there is a one in six chance that the other die will land on the same. For example, the probability of the first die being a four is 1/6, and the chances of a double four are one in thirty-six (since $1/6 \times 1/6 = 1/36$). So the chances of throwing any pair of numbers, a double, are of course one in six. And by multiplying 1/6 by 1/6 ten times, it is straightforward enough to calculate that the probability of throwing a double ten times in a row (regardless of what it is – for example, a double four, then a double one, and so on) is about one in sixty million! This means that if every person in Britain were to have a go at throwing a pair of dice ten times in succession then, statistically, only about one person will get all doubles every time.

But imagine that you were presented with a pair of dice that always lands on a double when thrown together. The actual number that they both land on appears to be random, usually changing at every throw, but both dice always end up rolling onto the same number. Clearly, you would assume some trickery. Perhaps these dice have some sophisticated internal mechanism that controls their motion, such that they land on numbers in an identical preprogrammed sequence? To test this theory you start by holding on to one of the dice while throwing the other, but thereafter throwing pairs of dice. Now any preprogrammed series will be out of step, so the trick shouldn't work. But despite this stratagem, the dice persist in landing on the same number.

Another possible explanation is that the dice must somehow be able to resynchronize before each throw by exchanging a remote signal. While such a mechanism seems rather sophisticated, it is at least possible to imagine in principle. However, any such mechanism would be subject to a limitation imposed by Einstein's theory of relativity, according to which no signal can travel faster than the speed of light. This provides you with a means of testing whether any signal is passing between the dice: all you need to do is ensure the dice are sufficiently far apart that there isn't enough time for any synchronizing signal to be exchanged in between throws. So let's imagine you try the same trick, as above, but somehow arrange for one die to be thrown on earth and the other to be thrown simultaneously on Mars. Even at its closest distance from earth, light takes four minutes to travel between the two planets, so you know that any synchronizing signal must suffer a similar delay. To beat it, you simply arrange for the two dice to be thrown at intervals more frequent than this. This should prevent any signal from synchronizing the dice between throws. If they continue to fall on matching numbers, then there would seem to have to be an intimate connection between them that ignores Einstein's famous limitation.

Although the above experiment hasn't been performed with interplanetary dice, analogous experiments have been performed with quantum-entangled particles on earth, and the results show that separated particles can perform the same kind of trick that we imagined for our dice: their state can remain correlated irrespective of the distance between them. This bizarre feature of the quantum world seems not to respect Einstein's cosmic speed limit, for a particle in one place can *instantaneously* influence another, however far apart the two may be. The term 'entanglement' to describe this phenomenon was coined by Schrödinger who, along with Einstein, was not a fan of what Einstein referred to as 'spooky action at a distance'. But, despite their scepticism, quantum entanglement has been proved in many experiments and is one of the most fundamental ideas in quantum mechanics, with many applications and examples in physics and chemistry – and, as we shall see, possibly in biology too.

To understand how quantum entanglement gets tangled up with biology we have to combine two ideas. The first is this instantaneous connection between two particles across space: entanglement. The second is that ability of a single quantum particle to be in a superposition of two or more different states at once: for example, an electron could be spinning both ways at once, so we would say it was in a superposition of 'spin up' and 'spin down' states. We combine these two ideas by having two entangled electrons in an atom, each in a superposition of its two spin states. Although neither has a definite spin direction, whatever it is doing influences and is influenced by the spin of its partner. But remember that pairs of electrons in the same atom are always in a singlet state, which means that they have to have opposite spin at all times: one must be spin-up and the other must be spin-down. So although both electrons are in a superposition of being both up and down at the same time, in a peculiar quantum way they must, at all times, have opposite spin.

Now let's separate the two entangled electrons so they are no longer in the same atom. If we then decide to measure the spin state of one electron we will force it to *choose* which way it is spinning. Say we find that, after measurement, it is spin-up. Because the electrons were in an entangled singlet spin state, this means that the other electron must now be spin-down. But remember that, before measurement, both were in a superposition of spinning up and down. After measurement both have distinct states: one of them is up and the other is down. So the second electron has instantly and remotely changed its physical state from being in a superposition of spinning both ways at once to being spin-down – without being touched. All we have done is to measure the state of its partner. And in principle it doesn't matter how far away this second electron is – it could be on the other side of the universe and the effect would be the same: measuring just one of an entangled pair *immediately* collapses the superposition of the other, irrespective of how far away it is.

Here is a useful analogy that may help you (just a little!). Imagine a pair of gloves, each in a sealed box, but separated by many miles. You have in your possession one of the boxes and, before opening it, you do not know whether yours is the left-handed or right-handed glove. Once you open the box and discover the right-handed glove you instantly *know* that the other glove in the unopened box is left-handed, no matter how far away the other box is. What is crucial here, however, is that all that has changed is your knowledge. The remote box had always contained the left-handed glove, irrespective of whether or not you chose to open your box.

Quantum entanglement is different. Before the measurement, neither electron has a definite spin direction. It is only the act of measurement (of either entangled particle) that forces both electrons to change their state from each being in a quantum superposition of both up *and* down to being in a definite state of up *or* down; whereas with the gloves it was only your ignorance of the pre-existing definite state of the gloves that was banished. Not only does quantum measurement of one electron force it to 'choose' to spin either up or down; that 'choice' instantaneously forces its twin to adopt the complementary state, no matter how far away it is.

There is one further subtlety that needs to be added. As we have already discussed, two electrons are in a combined singlet state when they are coupled together and spinning in opposite directions, and in a triplet state when they are spinning in the same direction. If one electron from a singlet pair sitting in the same atom jumps across into a neighbouring atom, its spin can flip over so that it is now spinning in the same direction as the twin it left behind, creating a triplet spin state. However, despite now being in different atoms, the pair can still maintain their delicate entangled state in which they remain quantum mechanically coupled together.

But this is the quantum world, and just because the electron that jumped out of the atom *can* now flip its spin, this doesn't mean that it definitely *has*. Each of the two electrons will still be in a superposition of spinning both ways at once, and as such the pair will exist in a superposition of being in a singlet *and* a triplet state simultaneously: spinning in the same direction and in opposite directions at the same time!

So now that you have been suitably primed, and probably confused, it is time to introduce you to the strangest and yet most celebrated idea in the field of quantum biology.

A radical sense of direction

At the beginning of this chapter we discussed the problem of how something as weak as the earth's magnetic field can provide sufficient energy to alter the outcome of a chemical reaction and thereby generate a biological signal that will, for example, tell a robin in which direction it needs to fly. The Oxford-based chemist Peter Hore has a very nice analogy of how such extreme sensitivity might be possible:

> Imagine we have a block of granite weighing one kilogram and ask whether a fly could tip it over. Common sense says the answer is, surely, no. But suppose I were to poise the stone on one of its edges. Clearly it would not be stable in such a position and would tend to fall to the left or the right if left to its own devices. Now suppose that while the block is teetering in this way a fly were to land on its right hand side. Even though the energy imparted by the fly would be tiny, it could be enough to cause the block to fall to the right rather than the left.[18]

The moral is that tiny energies can have significant effects, but only if the system on which they operate is very finely balanced between two different outcomes. So, to detect the impact of the earth's very weak magnetic field we need the chemical equivalent of a granite block in a finely balanced state, such that it could be dramatically affected by the slightest of external influences, such as a weak magnetic field.

And now we come back to Klaus Schulten's fast triplet reaction. You may recall that electronic bonds between atoms are often formed by the sharing of a pair of electrons. This electron pair is always entangled and almost always in a singlet spin state: that is, the electrons have opposite spin. However, remarkably, the two electrons can remain entangled even after the bond between the atoms is broken. The separated atoms, which are now called *free radicals*, can drift apart, and it becomes possible for the spin of one of the electrons to flip over so the entangled electrons – now on different atoms – find themselves in a superposition of both singlet and triplet states, as in Schulten's fast triplet reaction.

An important feature of this quantum superposition is that it isn't necessarily equally balanced: the probabilities of our catching the entangled pair of electrons in the singlet or triplet state are not equal. And, crucially, the balance between these two probabilities is sensitive to any external magnetic field. In fact, the angle of the magnetic field with respect to the orientation of the separated pair strongly influences the likelihood of catching it in a singlet or triplet state.

Radical pairs tend to be very unstable, so their electrons will often recombine to form the products of the chemical reaction. But the precise chemical nature of the products will then depend on this singlet–triplet balance, with all its sensitivity to magnetic fields. To understand how it works, we can think of the free radical intermediate stage of the reaction state as being like that metaphorical balanced granite block. In this state, the reaction is so delicately poised that even a weak magnetic field – taking the place of the fly – of less than 100 microtesla, such as the earth's, is sufficient to influence the way that the singlet/triplet state coin toss *falls* to generate the products of the chemical reaction.[19] Here at last was a mechanism by which magnetic fields could influence chemical reactions, and thereby, Schulten claimed, provide a magnetic compass for birds.

But Schulten had no idea where in the bird's body this proposed radical pair reaction was taking place – presumably it would make most sense for it to be located in the brain. But for it to work, the radical pair had to be created in the first place (just as the granite block needs to be tipped onto its edge). He presented his work at Harvard in 1978, describing the experiments carried out by his group in Göttingen in which a laser pulse was used to create an entangled radical pair of electrons. In the audience was an eminent scientist named Dudley Herschbach, who would later go on to win a Nobel Prize in chemistry. At the end of the lecture Herschbach asked as a good-natured jibe: 'But Klaus, where in the bird is the laser?' Under pressure to provide a sensible answer to such a senior professor, Schulten suggested that if indeed light was needed to activate the radical pair, then maybe that process took place within the bird's eye.

In 1977, a year before Schulten's radical pair paper, an Oxford physicist named Mike Leask had speculated in another *Nature* paper that the origin of the magnetic sense might indeed lie within photo-receptors in the eye.[20] He had even suggested that the eye pigment molecule, rhodopsin, was responsible. When Wolfgang Wiltschko read Leask's paper he was intrigued, although he had no experimental evidence to suggest that light played a role in avian magnetoreception. So he set out to test Leask's idea.

At the time, Wiltschko had been conducting experiments on homing pigeons to see whether they gathered magnetic navigational information on their outward journey that they then used to find their way back home. He had found that subjecting the pigeons to a disrupting magnetic field while being transported away from their home messed up their ability to find their way back when released. Inspired by Leask's theory, he decided to conduct the experiment again, this time without the magnetic field disturbance. Instead, he transported the pigeons in total darkness in a box on the roof of his Volkswagen bus. The birds then had difficulty finding their way home, demonstrating that they required light to help them plot out a magnetic map of their outward journey, which they would then use to track their way home.

The Wiltschkos finally met Klaus Schulten at a conference in the French Alps in 1986. They were by this time convinced that the robin's magnetoreception relied on light entering its eye but, like almost everyone else interested in the biochemical effects of magnetic fields, they were not yet persuaded that the radical pair hypothesis was correct. Indeed, no one knew where in the eye the radical pair might form. Then, in 1998, the pigment protein cryptochrome was discovered in the eyes of fruit flies and, as we described earlier in the chapter, was shown to be responsible for the light-driven entrainment of their circadian rhythms. Crucially, cryptochrome was known to be the kind of protein capable of forming free radicals during its interaction with light. This was seized upon by Schulten and his co-workers to propose that cryptochrome was the elusive receptor for the avian chemical compass. Their work was published in 2000 and would become one of the classic papers of quantum biology.[21] The lead author on that paper was of course Thorsten Ritz, whom we also met in chapter 1 and who at this point was working on his PhD with Klaus Schulten. Now at the physics department at the University of California, Irvine, Thorsten is today regarded as one of the world's leading experts on magnetoreception.

The 2000 paper is important for two reasons. First, it proposed cryptochrome as the candidate molecule for the chemical compass; and second, it described in beautiful – albeit speculative – detail just how the bird's orientation in the earth's magnetic field might affect what it sees.

The first step in their scheme is the absorption of a photon of blue light by the light-sensitive pigment molecule, FAD, that sits within the cryptochrome protein, and which we met earlier in the chapter. As we described, the energy of this photon is used to eject an electron from one of the atoms within the FAD molecule, leaving behind an electron vacancy. This can be filled by another electron donated from an entangled pair of electrons in an amino acid called tryptophan within the cryptochrome protein. Crucially, however, the donated electron can remain entangled with its partner. The pair of entangled electrons can then form a superposition of singlet/triplet states, which is the chemical system that Klaus Schulten found to be so exquisitely sensitive to a magnetic field. Once again, the delicate balance between the singlet/triplet states is highly sensitive to the strength and angle of the earth's magnetic field, so that the direction in which the bird flies makes a difference to the composition of the final chemical products that are generated by the chemical reaction. Somehow, in a mechanism that isn't at all clear even now, this difference – which way the granite block tumbles – generates a signal that is sent to the bird's brain to tell it where the nearest magnetic pole lies.

This radical pair mechanism proposed by Ritz and Schulten was certainly very elegant; but was it real? At the time there wasn't even any evidence that cryptochrome can generate free radicals when exposed to light. However, in 2007 another German group, this time based at the University of Oldenburg and led by Henrik Mouritsen, were able to isolate cryptochrome molecules from the retina of the garden warbler and show that they did indeed produce long-lived radical pairs when exposed to blue light.[22]

We have no idea what this magnetic 'seeing' looks like to birds, but since cryptochrome is an eye pigment that is potentially doing a similar job to the opsin and rhodopsin pigments that provide colour vision, perhaps the birds' view of the sky is imbued with an extra colour invisible to the rest of us (just as some insects can see ultraviolet light) which maps onto the earth's magnetic field.

When Thorsten Ritz proposed his theory in 2000 there was no evidence for cryptochrome being involved in magnetoreception; but now, thanks to the work of Steve Reppert and colleagues, the same pigment is known to be involved in how fruit flies and monarch butterflies detect external magnetic fields. In 2004, researchers found three types of cryptochrome molecules in eyes of robins; and then in 2013 a paper from the Wiltschkos (still as active as ever, even though Wolfgang has now retired) demonstrated that cryptochrome extracted from the eyes of chickens* absorbed light at the same frequencies as those they discovered were important for magnetoreception.[23]

But does the process definitely rely on quantum mechanics in order to work? In 2004, Thorsten Ritz went to work with the Wiltschkos to try to differentiate between a conventional magnetite compass and a chemical compass based on their free radical mechanism. Compasses can of course be disrupted by anything magnetic: hold a compass close to a magnet and it will point to the magnet's north pole rather than the earth's. A standard bar magnet produces what is called a static magnetic field, which means that it doesn't change with time. However, it is also possible to generate an oscillating magnetic field – by, for example, rotating a bar magnet – and this is where it gets interesting. A conventional compass may still be disrupted by an oscillating magnetic field, but only if the oscillations are slow enough for the needle of the compass to track. If the oscillations are taking place very fast, say hundreds of times a second, then the needle of the compass can't track them any more, and their influence averages to zero. So a conventional compass may be disrupted by magnetic fields oscillating at low frequencies but not at high frequencies.

But a chemical compass will have a very different response. You will remember that the chemical compass was proposed to depend on radical pairs being in a superposition of singlet and triplet states. Because the two states differ in their energy and energy is related to frequency, the system will be associated with a frequency that, considering the energies involved, would be expected to be in the millions of oscillations per second range. A classical way of thinking about what is going on that may be easier to imagine (though it is not strictly correct) is that the entangled pair of electrons is flipping between singlet and triplet state many millions of times a second. In this state, the system can interact with an oscillating magnetic field

---

*Chickens do not of course migrate, even in the wild. But they still appear to retain magnetoreception.

by the process of resonance, but only if the field is oscillating at the same frequency as the radical pair: only if, to use our previous musical analogy, they are *in tune*. The resonance will then pump energy into the system that will change that critical balance between singlet and triplet states on which the chemical compass depends – essentially, tipping over that metaphorical granite block before it has time to detect the earth's magnetic field. So, in contrast to a conventional magnetite compass, a radical pair compass will be disrupted by magnetic fields that oscillate at very high frequencies.

The Ritz–Wiltschko team set up an experiment to test this very clear prediction of the radical pair theory using the European robin: would its compass be sensitive to low- or high-oscillating magnetic fields? They waited until the autumn, when the birds would be getting impatient to migrate south, and then placed them inside Emlen funnel chambers. They applied oscillating fields from various directions and at various frequencies and waited to see whether the fields could disrupt the birds' natural ability to orientate themselves.

The results were astonishing: a magnetic field tuned to 1.3 MHz (that is, oscillating at 1.3 million cycles per second), thousands of times weaker than even the earth's field, could nevertheless disrupt the birds' ability to orientate themselves. But increasing or decreasing the frequency of the field made it less effective. So the field appeared to be resonating with something vibrating at very high frequencies in the avian compass: clearly not a conventional magnetite-based compass, but something consistent with an entangled radical pair in a superposition of singlet and triplet states. This intriguing result[24] also shows that, if it exists, the entangled pair must be able to survive in the face of decoherence for at least a microsecond (a millionth of a second), because otherwise its lifetime would be too short to experience the ups and downs of the applied oscillating magnetic field.

However, the significance of this result has recently been questioned. Henrik Mouritsen's group at the University of Oldenburg showed that manmade electromagnetic noise, from a wide range of electronic devices, seeping through the walls of the unscreened wooden huts housing the birds at the university campus disrupted their magnetic compass orientation. But the capability returned once they were placed in aluminium-screened huts, which cut out about 99 per cent of the urban electromagnetic noise. Crucially, their results suggest that the disruptive effect of radio-frequency electromagnetic fields may not be confined to a narrow frequency band after all.[25]

So there are still aspects of the system that remain mysterious; for example, why the robin's compass should be so hypersensitive to oscillating magnetic fields, and how free radicals can remain entangled for long enough to make a biological difference. But in 2011, a paper from Vlatko Vedral's laboratory in Oxford presented quantum theoretical calculations of the proposed radical pair compass and demonstrated that superposition and entanglement should be sustained for at least tens of microseconds, greatly exceeding the durations achieved in many comparable manmade molecular systems; and potentially long enough to tell a robin which way it needs to fly.[26]

These remarkable studies have sparked an explosion of interest in magnetoreception, which has now been demonstrated in a wide range of species including a whole host of bird species, spiny lobsters, stingrays, sharks, fin whales, dolphins, bees and even microbes. In most cases, the mechanisms involved haven't yet been investigated, but cryptochrome-associated magnetoreception has now been discovered in a wide range of creatures from our doughty robin to the chickens and fruit flies we have already mentioned and several other organisms, including plants.[27] A study published by a Czech group in 2009 demonstrated magnetoreception in the American cockroach and showed that, as with the European robin, it was disrupted by high-frequency oscillating magnetic fields.[28] A follow-up study presented at a conference in 2011 showed that the cockroaches' compass required functional cryptochrome.

The discovery of a capability and a shared mechanism so widely distributed in nature suggests that it has been inherited from a common ancestor. But the common ancestor of chickens, robins, fruit flies, plants and cockroaches lived way, way back: more than 500 million years ago. So quantum compasses are probably ancient, and are likely to have provided navigational skills for the reptiles and dinosaurs that roamed the Cretaceous swamps alongside the T. rex we met in chapter 3 (remember that modern birds such as robins are descended from dinosaurs), the fish that swam the Permian seas, the ancient arthropods that crawled over or burrowed beneath the Cambrian oceans, and maybe even the pre-Cambrian microbes that were the ancestors of all cellular life. It seems that Einstein's spooky action at a distance may have been helping creatures to find their way around the globe for most of the history of our planet.

7

Quantum genes

THE COLDEST place on earth is not, as you might imagine, the south pole, but somewhere in the middle of the east Antarctic ice sheet, nearly 1,300 kilometres from the pole. There, winter temperatures routinely plummet to many tens of degrees Celsius below zero. The lowest temperature ever measured on earth, −82.9°C, was recorded there on 21 July 1983, earning the region the title of 'the Southern Pole of Cold'. In temperatures this low, steel shatters and diesel fuel has to be cut with a chainsaw.

The extreme cold freezes any moisture out of the air, which, together with the strong winds that blow unceasingly across the frozen plains, probably makes the east Antarctic the most inhospitable place on the planet.

But it wasn't always such a hostile place. The landmass that forms Antarctica was once part of the supercontinent known as Gondwanaland and was in fact located near the equator. It was covered by a thick vegetation of seed ferns, ginkgo trees and cycads that were grazed on in turn by dinosaurs and herbivorous reptiles, such as the rhino-like Lystrosaurus. But about eighty million years ago the landmass started to break up and a fragment drifted southward, eventually settling over the south pole to become Antarctica. Then, about sixty-five million years ago, a massive asteroid hit the earth, wiping out all the dinosaurs and giant reptiles and leaving ecological space for warm-blooded mammals to become dominant. Despite being very far from the impact site, Antarctica's fauna and flora were radically altered as ferns and cycads were replaced by deciduous forests. These were inhabited by now extinct marsupials, reptiles and birds, including giant penguins. Fast-flowing rivers, and deep lakes teeming with bony fish and arthropods, filled the valleys.

But as greenhouse gas levels dropped, so did the temperature in Antarctica. Circulating ocean currents encouraged further cooling, and about thirty-four million years ago the surface waters of the rivers and inland lakes started to freeze in the winter. Then about fifteen million years ago the winter ice finally failed to melt in the summer, locking the lakes and rivers beneath a solid frozen roof. As our planet continued to cool, massive glaciers marched over Antarctica, extinguishing all its terrestrial mammals, reptiles and amphibians, and burying the land, lakes and rivers beneath gigantic sheets of ice several kilometres thick. Antarctica has remained locked in a deep freeze ever since.

It was only in the nineteenth century that the American sealer Captain John Davis became the first human known to have set foot on the continent; and only in the twentieth that permanent settlement began, as several countries raced to establish their territorial claims by building research stations on the continent. The first Soviet Antarctic station, Mirny, was established near the coast on 13 February 1956, and it was from here, two years later, that an expedition left for the interior of the continent with the aim of setting up a base at its geomagnetic pole. The expedition was dogged by snowstorms, loose snow, extreme cold (−55°C) and lack of oxygen, but finally arrived at the geomagnetic south pole on 16 December, during the southern hemisphere's summer, and established the Vostok station.

Since then, that research base has been manned nearly continuously with a team of between twelve and twenty-five scientists and engineers who make geomagnetic and atmospheric measurements. One of the main purposes of the station is to drill into the underlying glacier to capture a frozen record of past climates. In the 1970s the engineers drilled a set of cores up to 952 metres deep, reaching ice laid down in the last ice age, tens of thousands of years ago. New rigs arrived in the 1980s, allowing the researchers to reach a depth of 2,202 metres. By 1996 they had managed to drill down to 3,623 metres: a hole in the ice over two miles deep to a level laid down as surface ice 420,000 years ago.

But then the drilling was stopped, because something odd had been detected lying not far beneath the bottom of the borehole. In fact, the discovery that something unusual lay beneath the Vostok station had been made a couple of decades earlier, in 1974, when a British seismic survey of the region had revealed anomalous readings for a large area covering 10,000 square kilometres and lying about 4 kilometres below the ice. The Russian geographer Andrey Petro-vich Kapitsa suggested that the radar anomaly was caused by a huge lake trapped beneath the ice and kept warm enough to remain liquid by the extreme pressure and by underlying geothermal energy. Kapit-sa's proposal was eventually confirmed by satellite measurements of the area in 1996, which revealed a subglacial lake up to 500 metres deep (from the top of its liquid surface to its bottom) and the size of Lake Ontario. The team named it Lake Vostok.

With an ancient lake buried beneath the ice, the drilling opera-tions at the Vostok station took on a wholly different significance as the borehole approached a unique environment. Lake Vostok had been locked away from the earth's surface for hundreds of thousands, if not millions, of years* – a lost world. What had happened to all those animals, plants, algae and microbes that thrived in the lake before it was shut off, trapping any surviving organisms in absolute darkness and cold? Had all life been extinguished, or could some creatures have survived and even adapted to life several kilometres beneath the surface of the glacier? Such hardy organisms would have had to cope with an extreme environment: bitterly cold and totally dark, in water compressed by the weight of the thick ice sheet to

---

* The bottom of the glacier that sits on the lake today was laid down more than four hundred thousand years ago but the lake may have been frozen for a lot longer. It isn't clear whether the current glacier replaced earlier glaciers or the lake experienced ice-free periods between ice ages.

more than three hundred times the pressure of any surface lake. However, surprisingly diverse life does manage to eke out a living in other unlikely places, such as the scorching sulphurous edges of volcanoes, acid lakes, and even deep, dark submarine trenches thousands of metres below the ocean surface. Perhaps Vostok too could support its own ecosystem of *extremophiles*.*

The discovery of a lake under the deep ice acquired even greater significance thanks to another discovery nearly half a billion miles away in 1980, when the Voyager 2 spacecraft photographed the surface of Jupiter's moon Europa, revealing an icy surface with tell-tale signs of a liquid ocean lying beneath it. If life could survive for hundreds of thousands of years in waters buried kilometres beneath an Antarctic glacier, then maybe Europa's submerged oceans could support alien life. The search for life within Lake Vostok became a rehearsal for the even more thrilling hunt for life beyond our planet.

The drilling was halted in 1996, just 100 metres above the surface of the lake, to prevent its pristine waters from coming into contact with the kerosene-saturated drill bit, potentially contaminated with plants, animals, microbes and chemicals from the surface. However, Lake Vostok's water had already been studied from previously extracted ice cores. Thermal currents drive the water in the lake so that just beneath its icy ceiling it is going through a continual cycle of freezing and thawing. This process has continued ever since the lake was sealed off, so its roof is made up not of glacier ice, but of frozen lake water – known as *accretion* ice – that extends to tens of metres above the liquid surface of the lake. The cores extracted from the earlier drilling operations had penetrated down to this level of ice and, in 2013, the first detailed study of the Vostok accretion ice cores was published.[1] The conclusion of the work was that the ice-locked lake contains a complex web of organisms, including single-celled bacteria, fungi and protozoa, along with more complex animals such as molluscs, worms, anemones and even arthropods. Scientists have

---

* Organisms that live in extreme (from our perspective) environments.

even managed to identify what kind of metabolisms were used by these creatures, as well as their likely habitats and ecology.

What we want to focus on in this chapter is not the undeniably fascinating biology of Vostok, but the means by which any ecosystem could survive, locked away, for thousands or even millions of years. Indeed, Vostok can be considered to be a kind of microcosm of the earth itself, which has been virtually locked away from inputs, apart from solar photons, for four billion years and yet has maintained a rich and diverse ecosystem in the face of challenges from massive volcanic eruptions, asteroid impacts and climate shifts. How does the vast complexity of life manage to thrive and endure through extreme shifts in its environment for thousands or even millions of years?

A clue can be found in some of the material that was studied by the Vostok biology team: a few micrograms of a chemical extracted from frozen lake water. This chemical is crucial to the continuity and diversity of all life on our planet and contains the most extraordinary molecule in the known universe. We call it DNA.

The group that performed the Vostok DNA study are based at Bowling Green State University in the US. To read the sequence of millions of fragments of Vostok DNA molecules recovered from the lake water, they used the kind of DNA sequencing technology that had previously been used to decipher the human genome. They then compared the Vostok DNA to databases packed full of gene sequences read from the genomes of thousands of organisms collected from around the globe. What they discovered was that many of the Vostok sequences were identical or very close matches to genes from bacteria, fungi, arthropods and other creatures that live above the ice, particularly those inhabiting cold lakes and deep, dark marine trenches – environments that are probably a bit like Lake Vostok. These gene similarities allowed them to make educated guesses about the likely nature and habits of the kind of organisms that had left their DNA signatures under the ice.

But remember that the Vostok organisms have been locked under the ice for many hundreds of thousands of years. The similarity of their DNA sequences to those of organisms that live above the ice is thus a consequence of shared ancestry from organisms that must have lived among the flora and fauna of Antarctica before the lake and its inhabitants were locked away beneath the ice. The gene sequences of those ancestral organisms were then copied, independently, both above and below the ice, for thousands of generations. Yet despite this long chain of copying events, the twin versions of the same genes have remained nearly identical. Somehow, the complex genetic information that determines the shape, characteristics and function of the organisms that live both above and beneath the ice has been faithfully transmitted, with hardly any errors, over hundreds of thousands of years.

This ability of genetic information to replicate itself faithfully from one generation to the next – what we call heredity – is, of course, central to life. Genes, written into DNA, encode the proteins and enzymes that, via metabolism, make every biomolecule of every living cell, from the photosynthetic pigments of plants and microbes to the olfactory receptors of animals or the mysterious magnetic compasses of birds, and indeed every feature of every living organism. Many biologists would argue, indeed, that self-replication is life's defining feature. But living organisms could not replicate themselves unless they were capable of first replicating the instructions for making themselves. So the process of heredity – high-fidelity copying of genetic information – makes life possible. You may remember from chapter 2 that the mystery of heredity – how genetic information can be transmitted so faithfully from one generation to the next – was the puzzle that convinced Erwin Schrödinger that genes were quantum mechanical entities. But was he right? Do we need quantum mechanics to account for heredity? This is the question to which we will now return.

Fidelity

We tend to take for granted the ability of living organisms to replicate their genomes accurately, but it is in fact one of the most remarkable and essential aspects of life. The rate of copying errors in DNA replication, what we call mutations, is usually less than one in a billion. To get some idea of this extraordinary level of accuracy, consider the one million or so letters, punctuation marks and spaces in this book. Now consider one thousand similarly sized books in a library and imagine you had the job of faithfully copying every single character and space. How many errors do you think you would make? This was precisely the task performed by medieval scribes, who did their best to hand-copy texts before the invention of the printing press. Their efforts were, not surprisingly, riddled with errors, as shown by the variety of divergent copies of medieval texts. Of course, computers are able to copy information with a very high degree of fidelity, but they do so with the hard edges of modern electronic digital technology. Imagine building a copying machine out of wet, squishy material. How many errors do you think it would make in reading and writing its copied information? Yet when that wet squishy material is one of the cells in your body and the information is encoded in DNA then the number of errors is less than one in a billion.

High-fidelity copying is crucial for life because the extraordinary complexity of living tissue requires an equally complex instruction set, in which a single error may be fatal. The genome in our cells consists of about three billion genetic letters which encode about fifteen thousand genes, but the genomes of even the simplest self-replicating microbes, such as those that live under the Vostok ice, consist of several thousand genes written into several million genetic letters. Although most organisms tolerate a few mutations at each generation, allowing more than a handful into the next generation can lead to severe problems, which we humans experience as genetic diseases, or even non-viable offspring. Also, whenever the cells in our body – blood cells, skin cells, etc. – replicate, they must also replicate their DNA to insert into daughter cells. Errors in this process lead to cancer.*

But to understand how quantum mechanics is central to heredity we must first visit Cambridge in 1953 where, on 28 February, Francis Crick rushed into the Eagle pub and declared that he and James Watson had 'discovered the secret of life'. Later that year they published their landmark paper,[2] which unveiled a structure and described a set of simple rules that provided the answers to two of the most fundamental mysteries of life: how biological information is encoded and how it is inherited.

What tends to be emphasized in many accounts of the discovery of the genetic code is arguably a feature of secondary significance: that DNA adopts a double-helical structure. This is indeed remarkable, and the elegant structure of DNA has rightly become one of the most iconic images in science, reproduced on T-shirts and websites and even in architecture. But the double helix is essentially just a scaffold. The real secret of DNA lies in what the helix supports.

As we outlined briefly in chapter 2, the helical structure of DNA (figure 7.1) is provided by a sugar–phosphate backbone that carries the actual message of DNA: the strings of nucleic acid bases, guanine (G), cytosine (C), thymine (T) and adenine (A). Watson and Crick recognized that this linear sequence formed a code – and this, they proposed, was the genetic code.

In the last line of their historic paper, Watson and Crick suggested that the structure of DNA also provided a solution to the second of life's great mysteries: 'It has not escaped our notice that the specific pairing we have postulated immediately suggests a possible copying mechanism for the genetic material.' What hadn't escaped their notice was a crucial feature of the double helix, that the information on one of its strands – its sequence of bases – is also present

---

* Cancers are caused by mutations in genes that control cell growth, leading to uncontrolled cell growth and thereby tumours.

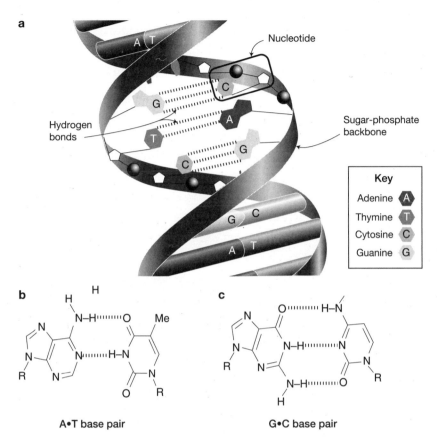

b

A•T base pair

c

G•C base pair

Figure 7.1: The structure of DNA: (a) shows Watson and Crick's double helix; (b) shows a close-up of the paired genetic letters A and T; (c) shows a close-up of the paired genetic letters G and C. In both instances, the hydrogen bonds – shared protons – that link the two bases are indicated as dotted lines. In this standard (canonical) Watson and Crick base pairing the bases are in their normal, non-tautomeric form.

as an inverse copy on the other strand: an A on one strand is always paired with a T on the other and a G is always paired with a C. The specific pairing between the bases on opposite strands (an A:T pair or a G:C pair) is actually provided by weak chemical bonds, called hydrogen bonds. This 'glue' holding two molecules together is essentially a shared proton and is central to our story, so we will be considering its nature in more detail shortly. But the weakness of the bonding between the paired DNA strands immediately suggested a copying mechanism: the strands could be pulled apart and each could act as a template on which to build its complementary partner to make two copies of the original double-strand. This is precisely what happens when genes are copied during cell division. The two strands of the double helix with their complementary information are pulled apart to allow an enzyme called *DNA polymerase* access to each separated strand. The enzyme then attaches to a single strand and slides along the chain of nucleotides, reading each genetic letter and, with almost unerring accuracy, inserting a complementary base into the growing strand: whenever it sees an A it inserts a T, whenever it sees a G it inserts a C, and so on until it has made a complete complementary copy. The same process is repeated on the other strand, giving rise to two copies of the original double helix: one for each daughter cell.

This deceptively simple process underpins the propagation of all life on our planet. But when Schrödinger insisted in 1944 that the extraordinarily high degree of fidelity of heredity could not be accounted for by classical laws – genes, he insisted, were just too small for their regularity to be based on the 'order from disorder' rules – he proposed that genes must instead be some kind of *aperiodic crystal*. Are genes aperiodic crystals?

Crystals, such as salt grains, tend to have distinctive shapes. Sodium chloride (common salt) crystals are cubes, whereas water molecules in ice form hexagonal prisms that grow into the marvellously diverse forms of snowflakes. These shapes are a consequence of the ways molecules can pack together inside the crystal, so, ultimately, they are determined by the quantum laws that determine the shape of molecules. But standard crystals, although highly ordered, don't encode much information because each repeated unit is the same as all the others – a bit like a tessellated wallpaper pattern – so a simple rule can describe the entire crystal. Schrödinger proposed that genes were what he called aperiodic crystals: that is, crystals with a similar repeated molecular structure to standard crystals, but modulated in some way, for example with different intervals or periods (hence 'aperiodic') between the repeats or different structures in the repeats – more like complex tapestry than wallpaper. He proposed that these modulated repeated structures encode genetic information, and that, like crystals, their order would be encoded at the quantum level. Remember that this was a decade before Watson and Crick: years before the structure of a gene, or even what genes were made of, was known.

Was Schrödinger right? The first obvious point is that the DNA code is indeed made of a repeated structure – the DNA bases – that is aperiodic in the sense that each repeating unit can be occupied by one of four different bases. Genes really are aperiodic crystals, just as Schrödinger predicted. But aperiodic crystals don't necessarily encode information at the quantum level: the irregular grains on a photographic plate are made of silver salt crystals, and they aren't quantum. To see if Schrödinger was also right about genes being quantum entities we need to take a deeper look at the structure of DNA bases, and in particular at the nature of the complementary base-pair bonding, T to A and C to G.

The DNA pairing that holds the genetic code is rooted in the chemical bonds that hold the complementary bases together. As we have already mentioned, these bonds, called hydrogen bonds, are formed by single protons, essentially nuclei of hydrogen atoms, which are shared between two atoms, one in each of the complementary bases on opposite strands: it is these that hold the paired bases together (figure 7.1). Base A has to pair with base T because each A holds protons at precisely the right positions to form hydrogen bonds with a T. An A base cannot pair with a C base because the protons would not sit in the right places to make the bonds.

This proton-mediated pairing of nucleotide bases *is* the genetic code that is replicated and passed on at each generation. And this isn't just a one-off transfer of information – like a coded message written on a 'one-time' pad which is destroyed after use. The genetic code has to be continuously read throughout the life of the cell to direct the protein-making machinery to make the engines of life, enzymes, and thereby orchestrate all the other activities of the cell. This process is performed by an enzyme called *RNA polymerase* which, like DNA polymerase, reads the positions of those coding protons along the DNA chain. Just as the meaning of a message or the plot of a book is written into the position of letters on a page, so the positions of protons on the double helix determine the story of life.

The Swedish physicist Per-Olov Löwdin was the first to point out what seems obvious in hindsight: that the protons' position is determined by quantum, not classical, laws. So the genetic code that makes life possible is inevitably a quantum code. Schrödinger was right: genes are written in quantum letters, and the fidelity of heredity is provided by quantum rather than classical laws. Just as the shape of a crystal is determined ultimately by quantum laws, so the shape of your nose, the colour of your eyes and aspects of your character are determined by quantum laws operating within the structure of a single molecule of DNA that you inherited from one or other of your parents. As Schrödinger predicted, life works via order that goes all the way down from the structure and behaviour of whole organisms to the position of protons along its DNA strands – order from order – and it is this order that is responsible for the fidelity of heredity.

But even quantum replicators make the occasional mistake.

Infidelity

Life couldn't have evolved on our planet and adapted to its many challenges if the process of copying the genetic code was always perfect. For example, the microbes swimming in those temperate Antarctic lakes many thousands of years ago would have been well adapted to life in a relatively warm and bright environment. When the ice roof sealed their world, those microbes that copied their genomes with 100 per cent fidelity would almost certainly have perished. But many microbes made a few mistakes in the copying process and generated mutant daughters slightly different from themselves. Those daughters whose differences better equipped them to survive in a colder, darker environment would have thrived and gradually, over thousands of not-quite-perfect copying events, the descendants of those trapped microbes would have become well adapted to life in the submerged lake.

Once again, this process of adaptation through mutation (DNA replication errors) within Lake Vostok is a microcosm of the process that has been taking place around the globe for billions of years. The earth has suffered many major catastrophes throughout its long history, from huge volcanic eruptions to ice ages and meteor impacts. Life would have perished if it hadn't adapted to change via copying errors. Just as importantly, mutations have also been the driver of the genetic changes which turned the simple microbes that first evolved on our planet into the hugely diverse biosphere of today. A little infidelity goes a long way, given sufficient time.

As well as proposing that quantum mechanics was the source of the fidelity of heredity, Erwin Schrödinger made another bold suggestion in his 1944 book, *What is Life?*. He speculated that mutations may represent some kind of quantum jump within the gene. Is this plausible? To answer this question we need first to explore a controversy that goes to the heart of evolutionary theory.

The giraffe, the bean and the fruit fly

It is often stated that evolution was 'discovered' by Charles Darwin, but the fact that organisms have changed over geological time had been familiar to naturalists for at least a century before Darwin through the study of fossils. Indeed, Charles's grandfather, Erasmus Darwin, had been a keen evolutionist. But probably the most famous pre-Darwinian evolutionary theory was put forward by a French aristocrat with the impressive title of Jean-Baptiste Pierre Antoine de Monet, Chevalier de Lamarck.

Born in 1744, Lamarck was trained as a Jesuit priest, but on his father's death inherited just enough money to buy a horse on which he rode off to become a soldier and fight in the Pomeranian War against Prussia. His soldiering career was cut short when he was wounded, and he returned to Paris to work as a bank clerk, while studying botany and medicine in his spare time. He eventually found a job as a botanical assistant at the Jardin du Roi (the King's Garden), until the revolution removed the head of his employer. But Lamarck thrived in post-revolutionary France, gaining a chair at the University of Paris, where he switched the focus of his studies from plants to invertebrates.

Lamarck is one of the most underappreciated of the great scientists, at least in the Anglo-Saxon world. As well as coining the term 'biology' (from the Greek *bios*, life) he came up with a theory of evolution that did at least provide a plausible mechanism for evolutionary change, half a century before Darwin. Lamarck pointed out that organisms are able to modify their bodies in response to the environment during their lifetimes. For example, farmers accustomed to hard physical toil generally develop more muscular bodies than bank clerks. Lamarck then claimed that these acquired changes could be inherited by offspring and descendants, and thereby drive evolutionary change. His most famous and most mocked example is that of the imaginary antelope that stretched its neck to graze on the highest leaves in the tree. Lamarck proposed that the antelope's descendants inherited the acquired characteristic of the elongated neck and their progeny went through the same process until they eventually evolved into giraffes.

The Lamarckian theory of inherited adaptive change was generally ridiculed in the Anglo-Saxon world, as there was abundant evidence that characteristics acquired during an animal's lifetime were not generally inherited. For example, fair-skinned northern Europeans who migrated to Australia several hundred years ago are generally suntanned if they spend a lot of time outdoors but, out of the sun, their children will be just as pale as their ancestors. The adaptive change in response to strong sunshine, a suntan, is clearly not inherited. So, after the publication of *On the Origin of Species* in

1859, Lamarckian evolutionary theory was eclipsed by Darwin's theory of natural selection.*

It is Darwin's version of evolution that is emphasized today – the notion of the survival of the fittest, with an unforgiving nature honing the well adapted from its less perfect progeny. But natural selection is only half the story of evolution. For evolution to be successful, natural selection needs a source of variation on which to cut its teeth. This was a great puzzle for Darwin because, as we have already discovered, heredity is characterized by a remarkably high degree of fidelity. This may not be immediately apparent in sexual organisms that appear to be different from their parents, but sexual reproduction only reshuffles existing parental traits to generate offspring. In fact, in the early nineteenth century it was generally believed that the mixing of traits in sexual reproduction proceeds rather like the mixing of paint. If you take several hundred tins of paint of varied colours and mix half a tin of one with half a tin of another and repeat this process thousands of times, then you will eventually end up with several hundred tins of grey paint: the individual variation will be blended towards a population average. But Darwin needed variation to be continually maintained and indeed added to, if it was to be the source of evolutionary change.

Darwin believed that evolution proceeded very gradually by natural selection acting on tiny heritable variation:

> Natural selection can act only by the preservation and accumulation of infinitesimally small inherited modifications, each profitable to the preserved being; and as modern geology has almost banished such views as the excavation of a great valley by a single diluvial wave, so will natural selection, if it be a true principle, banish the belief of the continued creation of new organic beings, or of any great and sudden modification in their structure.[3]

* Of course, it could as easily be called Wallace's theory of natural selection, after the great British naturalist and geographer Alfred Russel Wallace who, during a bout of malarial fever while travelling in the tropics, came up with virtually the same idea as Darwin.

But the source of this raw material for evolution – the 'infinitesimally small inherited modifications' – was a great mystery. Oddities or 'sports' with heritable characteristics were well known to nineteenth-century biologists: for example, a sheep with extremely short legs was born on a New England farm in the late eighteenth century and was bred from to produce a short-legged variety called Ancon sheep that are easier to manage because they cannot jump fences. However, Darwin believed that these sports couldn't be the drivers of evolution because the changes involved were too big, generating often bizarre creatures that would be very unlikely to survive in the wild. Darwin had to find a source of smaller, less dramatic, heritable changes to provide the infinitesimally small variations needed for his theory to work. He never really resolved this problem in his lifetime. Indeed, in later editions of the *Origin of Species* he even resorted to a form of Lamarckian evolutionary theory to generate heritable minor variation.

Part of the solution had already been discovered during Darwin's lifetime by the Czech monk and plant breeder Gregor Mendel, whom we met in chapter 2. Mendel's experiments with peas demonstrated that small variations in pea shape or colour were indeed *stably* inherited: that is – crucially – these traits did not blend but bred true generation after generation, though often skipping generations if the character was recessive rather than dominant. Mendel proposed that discrete heritable 'factors', what we now call genes, encode biological traits and are the source of biological variation. So instead of seeing sexual reproduction in terms of tins of paint being mixed, think of pots of marbles of an immense variety of colours and patterns. Each mixing generation swaps half of the marbles from one pot with half from another pot. Crucially, even after thousands of generations, the individual marbles retain their distinct colours, just as traits may be transmitted without change for hundreds or thousands of generations. Genes thereby provide a stable source of variation on which natural selection can act.

Mendel's work was mostly ignored during his lifetime and

forgotten after it; so, as far as we know, Darwin was not aware of Mendel's theory of 'heritable factors' and its potential solution to the blending puzzle. So the problem of finding the source of the heritable changes that drive evolution led to a decline in support for Darwinian evolutionary theory towards the end of the nineteenth century. But as the century turned, Mendel's ideas were revived by several botanists studying plant hybridization who discovered laws governing the inheritance of variation. Like all good scientists who think they have found something new, they searched through the existing literature before publishing their results; and they were astonished to discover that their laws of inheritance had been described several decades earlier by Mendel.

The rediscovery of Mendelian factors, now renamed 'genes',* provided a solution to Darwin's blending puzzle, but they didn't immediately solve the problem of finding the source of novel genetic variation needed to drive long-term evolutionary change, since genes appeared to be inherited without alteration. Natural selection can act to change the mix of the gene marbles at each generation but, on its own, it doesn't make any new marbles. This impasse was broken by one of the botanists who rediscovered Mendelian genetics, Hugo de Vries, who was walking through a potato field when he spotted a completely novel variety of the evening primrose, *Oenothera lamarckiana*, taller than the usual plant and with oval-shaped petals rather than the familiar heart-shaped petals. He recognized this flower as a 'mutant'; and, more importantly, he showed that the mutant traits were passed on to the plant's progeny, so they were inherited.

The geneticist Thomas Hunt Morgan took the study of de Vries' mutations into the laboratory at Columbia University in the early 1900s, working with the ever-amenable fruit fly. He and his team

---

* The term 'genetics' was coined in 1905 by William Bateson, an English geneticist and a proponent of Mendel's ideas; the term 'gene' was suggested four years later by Danish botanist Wilhelm Johannsen to distinguish between the outward appearance of an individual (its phenotype) and its genes (its genotype).

exposed the flies to strong acids, X-rays and toxins in an effort to create mutants. Finally, in 1909, a fly emerged from its pupa with white eyes and the team demonstrated that, as with de Vries' oddly shaped primroses, the mutant trait bred like a Mendelian gene.

The marriage of Darwinian natural selection with Mendelian genetics and mutation theory eventually led to what is often known as the neo-Darwinian synthesis. Mutation was understood to be the ultimate source of heritable genetic variations that are mostly of little effect and sometimes even harmful, but occasionally make mutants fitter than their parents. The process of natural selection then kicks in to weed out less-fit mutants from a population while allowing the more successful variants to survive and proliferate. Eventually, fitter mutants become the norm and evolution proceeds by 'the preservation and accumulation of infinitesimally small inherited modifications'.

A key component of the neo-Darwinian synthesis is the principle that mutations occur randomly; variation is not generated in response to an evolutionary change. So when the environment changes, a species has to wait for the right mutation to come along – through random processes – in order to track that change. This is in contrast to the Lamarckian idea of evolution, which proposed instead that heritable adaptation – the giraffe's longer neck – arises in response to an environmental challenge and was thereafter inherited.

In the early twentieth century it wasn't yet clear whether heritable mutations occurred randomly, as the neo-Darwinians believed, or were generated in response to environmental challenges, as the Lamarckians believed. Remember that Morgan treated his flies with noxious chemicals or radiation to generate mutations. Perhaps, in response to these environmental challenges, the flies generated novel variations that helped them survive the environmental challenge. Like Lamarck's giraffe, they might have metaphorically *stretched their necks*, and then passed this adaptive trait on to their descendants as a heritable mutation.

Classic experiments performed by Salvador Luria, James Watson's PhD supervisor, and Max Delbrück at the University of Indiana in 1943 set out to test the rival theories. By this time bacteria had replaced fruit flies as the favoured subjects of evolutionary studies because of their ease of growth in the laboratory and fast generation times. It was known that bacteria could be infected with viruses, but if repeatedly exposed would rapidly evolve resistance by acquiring mutations. This offered an ideal situation in which to test the rival neo-Darwinian and Lamarckian theories of mutation. Luria and Delbrück set out to discover whether bacterial mutants able to resist viral infection already existed in the population, as predicted by neo-Darwinism, or arose only in response to an environmental challenge by a virus, as predicted by Lamarckism. The two scientists found that the mutants occurred at pretty much the same rate whether the virus was present or absent. To put it another way, the mutation rate was not affected by the selective pressure of the environment. Their experiments earned them a Nobel Prize in 1969 and established the principle of the randomness of mutation as a cornerstone of modern evolutionary biology.

Yet when Luria and Delbrück were performing their experiments in 1943, still no one knew what these gene marbles were made of, even less what the physical mechanisms were that were responsible for generating mutations – changing one marble into another. That all changed in 1953 when Watson and Crick unveiled the double helix. The gene marbles were shown to be made of DNA. The principle that mutations were random then made perfect sense, since well-established causes of mutation, such as radiation or mutagenic chemicals, would tend to damage the DNA molecule randomly along its entire length, causing mutations in whatever genes they affected, irrespective of whether or not the change provided an advantage.

In their second paper on the structure of DNA,[4] Watson and Crick suggested that a process called tautomerization, which involves the movement of protons within a molecule, could also be a cause of mutation. As I am sure you are well aware by now, any process that involves the movement of fundamental particles, like protons, can be quantum mechanical. So, was Schrödinger right? Are mutations a kind of quantum jump?

Coding with protons

Take another look at the bottom half of figure 7.1. You will see that we have drawn the hydrogen bond – which, remember, is a shared proton – as a dotted line between two atoms (oxygen, O or nitrogen, N) on the paired bases. But isn't a proton a particle? Why, then, is it drawn as a dotted line rather than a single dot? The reason is, of course, that protons are quantum entities that have both particle and wave character: so the proton is delocalized, behaving like a smeared-out entity or a wave that sloshes between the two bases. The position of the H in figure 7.1 – denoting the most likely position of the proton – is not halfway between the two bases, but rather is offset to one side: closer to either one strand or the other. This asymmetry is responsible for an extremely important feature of DNA.

Let's consider one possible base pair, such as A–T, with the A on one strand and the T on the other, held together by two hydrogen bonds (protons) where one proton is closer to a nitrogen atom in A and the other is closer to an oxygen atom in T (figure 7.2a), allowing the formation of the A:T hydrogen bond. But remember that 'closer than' is a slippery concept in the quantum world where particles don't have fixed positions but inhabit a range of probabilities of being in many different places at once, including those that can only be reached by tunnelling. If the two protons that hold the genetic letters together were each to jump to the other side of their respective hydrogen bonds, then they would each end up closer to the opposite base. This results in the formation of alternative forms of each base called *tautomers* (figure 7.2b). Each of the DNA bases can therefore exist both in its common canonical form, as seen in

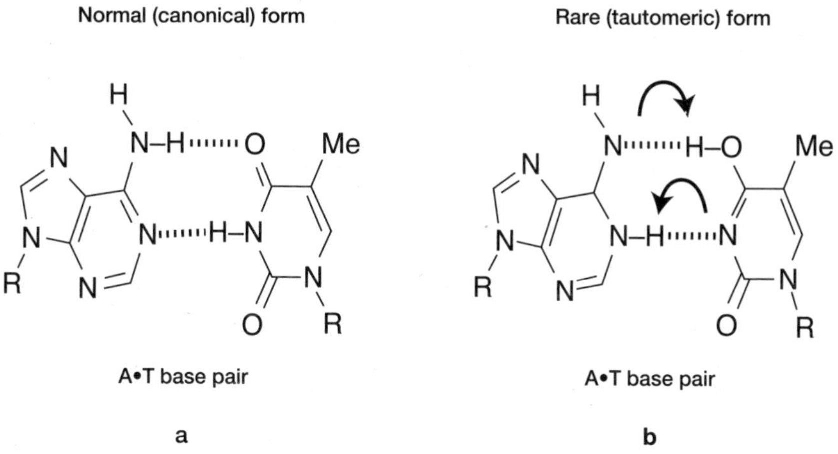

Normal (canonical) form

Rare (tautomeric) form

A•T base pair

a

A•T base pair

b

Figure 7.2: (a) A standard A–T base pair with the protons in their normal positions; (b) here the paired protons have jumped across the double helix to form the tautomeric form of both A and T.

Watson and Crick's double helix structure, and in the rarer tautomer, with its coding protons shifted across to new positions.

But remember that the protons forming the hydrogen bonds in DNA are responsible for the specificity of base-pairing that is used to replicate the genetic code. So, if the pair of coding protons move (in opposite directions), they are effectively rewriting the genetic code. For example, if a genetic letter in a DNA strand is a T (thymine) then in its normal form it pairs, correctly, with A. However, if a double proton swap occurs then both T and A will adopt their tautomeric forms. Of course, the protons may jump back again but if they happen to be in their rare tautomeric forms* at the time the DNA strand is being copied then the wrong bases may be incorporated into the new DNA strands. The tautomeric T can pair with G, rather than A, so G will be incorporated into the new strand where there was an A in the old strand. Similarly, if A is in its tautomeric state

---

*The alternative tautomeric forms of guanine and thymine are known as enol or keto, depending on the position of the coding protons; whereas cytosine and adenine tautomers are known as keto or amino forms.

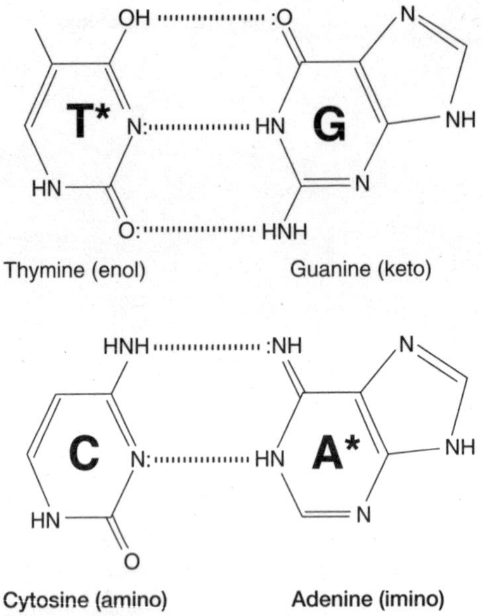
OH :O
N
T* N: HN G
NH
HN
N
O: HNH
Thymine (enol)
Guanine (keto)
HNH :NH
N
C N: HN A*
NH
HN
N
O
Cytosine (amino)
Adenine (imino)

Figure 7.3: In its tautomeric (enol) form, indicated by T* in the figure, T can pair incorrectly with G, rather than its usual partner, A. Similarly, the tautomeric form of A (A*) can pair incorrectly with C, rather than T. If these errors are incorporated during DNA replication then a mutation will result.

when the DNA is being replicated then it will pair with C, rather than T, so the new strand has C, where the old strand had T (figure 7.3). In either case, the newly formed DNA strands will carry mutations – changes in the DNA sequence that will be inherited by progeny.

Although this hypothesis is entirely plausible, it has been difficult to obtain direct evidence for it; but, in 2011, nearly sixty years after Watson and Crick published their paper, a group based at Duke University Medical Center in the US managed to demonstrate that incorrectly paired DNA bases with protons in the tautomeric position can indeed fit into the active site of DNA polymerase (the enzyme that makes new DNA), so are likely to be incorporated into newly replicated DNA to cause mutations.[5]

So tautomers – with alternative proton positions – appear to be a driver of mutation, and thereby of evolution; but what makes protons move to the wrong position? The obvious 'classical' possibility would be that they are occasionally 'shaken' across by the constant molecular vibrations going on all around them. However, this requires the availability of sufficient thermal energy to provide the impetus, the 'shake'. Just as in the enzyme-catalysed reactions discussed in chapter 3, the proton has to overcome quite a steep energy barrier to make the move. Alternatively, the protons may be knocked across by a collision with nearby water molecules; but there aren't many water molecules close to the coding protons in DNA to provide them with such a kick.

But there is another route – one that was found to play an important role in the way enzymes transfer electrons and protons. One of the consequences of the wave-like nature of subatomic particles such as electrons and protons is the possibility of quantum tunnelling. The fuzziness in the position of any particle allows it to *leak* through an energy barrier. We saw in chapter 3 how enzymes utilize quantum tunnelling of electrons and protons by bringing molecules close enough together for tunnelling to take place. A decade after Watson and Crick published their seminal paper, the Swedish physicist Per-Olov Löwdin, whom we met earlier in this chapter, proposed that quantum tunnelling could provide an alternative way for protons to move across hydrogen bonds to generate the tautomeric, mutagenic, forms of nucleotides.

It is important to emphasize that DNA mutations are caused by a variety of different mechanisms, including damage caused by chemicals, ultraviolet light, radioactive decay particles, even cosmic rays. All of these changes take place at a molecular level and so are bound to involve quantum mechanical processes. As yet, however, there is no indication that the weirder aspects of quantum mechanics play a role in these sources of mutations. But if quantum tunnelling is shown to be involved in the formation of DNA base tautomers, then quantum weirdness could be playing a role in the mutations that drive evolution.

However, tautomeric forms of DNA bases account for about 0.01 per cent of all natural DNA bases, potentially leading to errors of the same scale. This is a far higher rate than the one in a billion or so rate of mutation we find in nature, so if tautomeric bases are indeed present in the double helix then most of the resulting errors must be removed by the various error correction ('proofreading') processes that help to ensure the high fidelity of DNA replication. Even so, those errors promoted by quantum tunnelling that escape the correction machinery may be a source of the naturally occurring mutations that drive the evolution of all life on earth.

Discovering the underlying mechanisms of mutation is not only important for our understanding of evolution; it may also provide insight into how genetic diseases arise or how cells become cancerous, as both of these processes are caused by mutations. However, the problem with testing whether quantum tunnelling is involved is that, unlike other known causes of mutations such as chemical mutagens or radiation, it cannot be simply turned on or off. It is therefore not easy to measure mutation rates with and without tunnelling to see if they are different.

But there may be an alternative way of detecting a quantum mechanical origin to mutation, one that goes back to the difference between classical information and quantum information. Classical information can be read and reread over and over again without changing its message, whereas quantum systems are always perturbed by measurement. So when the DNA polymerase enzyme scans a DNA base to determine the position of coding protons, it is carrying out a quantum measurement, no different in principle from when a physicist measures the position of a proton in the laboratory. In both processes, the measurement is never innocuous: according to quantum mechanics, any measurement, whether performed by the DNA polymerase enzyme inside a cell or by a Geiger counter in a laboratory, inevitably changes the state of the particle being measured. If the state of that particle corresponds to a letter of the genetic code, then measurement, particularly frequent measurement, would be expected to change that code and potentially cause a mutation. Is there any evidence for this?

Although our entire genome is copied during DNA replication, most *readings* of our genes take place not during DNA replication but during the processes whereby genetic information is used to direct the synthesis of proteins. The first of these two processes, known as *transcription*, involves the copying of DNA-encoded information into RNA, a chemical cousin of DNA. The RNA then travels to the protein synthesis machinery to make proteins: this is the second process, known as *translation*. To distinguish these processes from the copying of genetic information during DNA replication we will refer to them as *reading* DNA.

A key characteristic of this process is that some genes are read much more often than others. If reading the DNA code during transcription constitutes a quantum measurement, then the more frequently read genes would be expected to be subject to more measurement-induced perturbations, leading to higher rates of mutation. This is indeed what is claimed to have been found in some studies. For example, Abhijit Datta and Sue Jinks-Robertson, from Emory University in Atlanta in the US, manipulated a single gene in yeast cells such that it was read either just a few times to make small quantities of protein in the cell or lots of times to make loads of protein. They discovered that the rate of mutation in the gene was thirty times higher when it was read at the higher levels.[6] A similar study in mouse cells found the same effect,[7] and a recent study of human genes concluded that those of our genes that are read at the highest levels tend to be mutated the most.[8] This is at least consistent with a quantum mechanical measurement effect, but of course it does not *prove* that quantum mechanics is involved. The reading of the DNA involves biochemical reactions that may disturb or damage the molecular structure of genes in many different ways, causing mutations, without any recourse to quantum mechanics.

To test whether quantum mechanics is involved in a biological process we require evidence that is hard or impossible to make sense of without quantum mechanics. In fact, it was a puzzle of this sort that first got the two of us interested in the role that quantum mechanics might play in biology.

Quantum jumping genes?

In September 1988, a paper on bacterial genetics written by a very eminent geneticist called John Cairns, working at the Harvard School of Public Health in Boston, was published in *Nature*.[9] The paper appeared to contradict that fundamental tenet of neo-Darwinian evolutionary theory: the principle that mutations, the source of genetic variation, occur randomly and that the direction of evolution is supplied by natural selection – the 'survival of the fittest'.

Cairns, an Oxford-educated British physician and scientist, worked in Australia and Uganda before taking a sabbatical in 1961 at the world-famous Cold Spring Harbor Laboratory in New York State. From 1963 to 1968 he served as director of the laboratory, which was a hotbed for the emerging science of molecular biology, particularly in the 1960s and 1970s when the scientists working there included figures such as Salvador Luria, Max Delbrück and James Watson. Cairns had actually met Watson many years earlier, when the rather dishevelled future Nobel laureate delivered a rambling presentation at a meeting in Oxford, and had not been hugely impressed; in fact, Cairns's overall impression of one of science's immortals was: 'I thought he was a complete nutter'.[10]

At Cold Spring Harbor Cairns carried out several landmark studies. For example, he demonstrated how DNA replication starts at a single point and then moves along the chromosome, rather like a train running on a track. He must also have eventually warmed to James Watson, because in 1966 they jointly edited a book on the role of bacterial viruses in the development of molecular biology. Then, in the 1990s, he took an interest in that earlier Nobel Prize-winning study by Luria and Delbrück that appeared to have proved that mutations occur randomly, before an organism is exposed to any environmental challenge. Cairns reckoned that there was a weakness in Luria and Delbrück's experimental design, through which they were supposed to have proved that bacterial mutants resistant to a virus are pre-existing in the population, rather than arising in response to the exposure to the virus.

Cairns pointed out that any bacteria that weren't already resistant to the virus wouldn't have had time to develop new mutations adaptively in response to the challenge because they would have been killed very quickly by the virus. He came up with an alternative experimental design that gave the bacteria a better chance of developing mutations in response to a challenge. Instead of looking for mutations that conferred resistance to a deadly virus, he instead starved the cells and looked for mutations that would allow bacteria to survive and grow. Like Luria and Delbrück, he saw that a few mutants managed to grow straight away, showing that they were pre-existing in the population; but, in contrast to the earlier study, he observed many more mutants appearing much later, apparently *in response to* starvation.

Cairns's result contradicted the well-established principle that mutations occurred randomly; his experiments appeared to demonstrate that mutations tended to occur when they were advantageous. The findings appeared to support the discredited Lamarckian theory of evolution – the starved bacteria weren't growing long necks but, just like Lamarck's imaginary antelope, they appeared to be responding to an environmental challenge by generating heritable modifications: mutations.

Cairns's experimental findings were soon confirmed by several other scientists. Yet the phenomenon had no explanation within contemporary genetics and molecular biology. There was simply no known mechanism that would allow a bacterium, or indeed any creature, to *choose* which genes to mutate and when. The finding also appeared to contradict what is sometimes called the central dogma of molecular biology: the principle that information flows only one way during transcription, from DNA out to proteins to the environment of a cell or organism. If Cairns's results were right, then cells must also be capable of reversing the flow of genetic information, allowing the environment to influence what is written in DNA.

The publication of Cairns' paper unleashed a storm of controversy and an avalanche of letters to *Nature* attempting to make sense of the finding. As a bacterial geneticist, Johnjoe was profoundly puzzled by the phenomenon of 'adaptive mutations', as they came to be known. At the time, he was reading a lay account of quantum mechanics, John Gribbin's popular *In Search of Schrödinger's Cat*,[11] and couldn't help pondering whether quantum mechanics, particularly that enigmatic process of quantum measurement, could provide an explanation of the Cairns result. Johnjoe was also familiar with Löwdin's claim that the genetic code is written in quantum letters; so, if Löwdin were right, the genome of Cairns' bacteria would have to be considered as a quantum system. And if that were true, then enquiring whether a mutation was present would constitute a quantum measurement. Could the perturbing influence of quantum measurement provide an explanation of Cairns's odd result? To explore this possibility we need to take a closer look at Cairns's experimental setup.

Cairns had introduced millions of cells of the gut bacterium *E. coli** onto the surface of a gel in dishes containing only lactose sugar as food. The particular strain of *E. coli* that Cairns used had an error in one of its genes that made it incapable of eating lactose, so the bacteria starved. But they didn't die; they just hung around on the surface of the gel. What surprised Cairns and caused all the controversy was that they didn't stay that way for long. After several days he observed colonies appearing on the surface of the gel. Each colony was composed of mutants descended from a single cell in which a mutation had corrected the error in the DNA code of the defective lactose-eating gene. The mutant colonies continued to appear over several days, until the plates eventually dried out.

According to standard evolutionary theory, as exemplified by the Luria–Delbrück experiment, evolution of the *E. coli* cell should have required the presence of pre-existing mutants in the population. A

---

* *Escherichia coli.*

few of these did indeed appear early on in the experiment, but they were far too few to account for the abundant lactose-eating colonies that quickly appeared several days later, *after* the bacteria were placed in the lactose environment (in which the mutations could provide an adaptive advantage to the cells – hence the term 'adaptive mutations').

Cairns ruled out trivial explanations of the phenomenon, such as a generally increased rate of mutation. He also demonstrated that adaptive mutations would occur only in environments where the mutation provided an advantage. Yet his results could not be accounted for by classical molecular biology: mutations should occur at the same rate irrespective of whether lactose was present or not. However, if, as Löwdin argued, genes are essentially quantum information systems, then the presence of lactose would potentially constitute a quantum measurement as it would reveal whether or not the cell's DNA had mutated: a quantum-level event dependent on the positions of single protons. Could quantum measurement account for the difference in mutation rates that Cairns observed?

Johnjoe decided to offer his ideas up for scrutiny in the Physics Department at the University of Surrey. Jim was in the audience and, although sceptical, was nevertheless intrigued. We decided to work together to investigate whether the idea had any quantum legs and eventually came up with a 'hand-wavy'* model that we proposed could account for adaptive mutations; this we published in the journal *Biosystems* in 1999.[12]

The model starts from the premise that protons can behave quantum mechanically; so, that those in the DNA of the starving *E. coli* cells will occasionally tunnel over into the tautomeric (mutagenic) position, and can just as easily tunnel back again to their original positions. Quantum mechanically, the system must be considered to be in a superposition of both states, tunnelled and not tunnelled, with the proton described by a wave function that is spread

---

* By which we mean one lacking a rigorous mathematical framework.

over both sites, but which is asymmetrical – giving a much greater probability of finding the proton in the non-mutated position. Here, there is no experimental measuring device or apparatus to record where the proton is; but the measurement process we discussed in chapter 4 is carried out by the surrounding environment. This is taking place all the time: for example, reading of DNA by the protein synthesis machinery forces the proton to 'make up its mind' on which side of the bond it is sitting – either in the normal (no growth) or in the tautomeric (growth) position; and mostly it will be found in the normal position.

Let us imagine Cairns' plate of *E. coli* cells as a box of coins, with each coin representing the proton in the key nucleotide base in the lactose utilization gene.* This proton can exist in one of two states: 'heads', corresponding to the normal, non-tautomeric position, or 'tails', corresponding to the rare tautomeric position. We start off with all the coins being heads-up, corresponding to the start of the experiment with the proton in the non-tautomeric position. But, quantum mechanically, the proton is always in a superposition of both normal and tautomeric positions, so our imaginary quantum coins will similarly be in a superposition of heads and tails, with most of the probability wave favouring the heads-up, normal state. But the proton position will eventually be measured by its surrounding environment within the cell, forcing it to *choose* where it is, which we can imagine as a kind of molecular coin toss, with overwhelming probability of throwing a head. The DNA may be occasionally copied,† but any new strand will encode only the genetic information that's there, which nearly always encodes only the defective enzyme – so the cell will continue to starve.

But remember that the coin represents a quantum particle, a

---

* In reality there will be more than one hydrogen bond holding the base pair together, but the argument holds equally well if we simplify the picture to just one.
† Starved and stressed cells may continue to attempt to copy their DNA, but the replication is likely to be aborted because of the limited resources available, so only short stretches corresponding to a few genes are made.

proton in the DNA strand; so even after measurement it is free to slip back into the quantum world to re-establish the original quantum superposition. So after our coin has been tossed and landed on heads, it will be tossed again, and again and again. Eventually, it will land on tails. In this state, the DNA may again be copied, but now it will make the active enzyme. In the absence of lactose, this will still not make any difference because, without lactose, the gene is useless. The cell will continue to starve.

However, if lactose is present then the situation will be very different, because the corrected gene made by the cell will allow the cell to consume lactose, grow and replicate. A return to the quantum superposition state will no longer be possible. The system will be irreversibly captured into the classical world as a mutant cell. We can conceive of this as – only in the presence of lactose – taking those rare coins that fall on tails out of the box and placing them in another box, marked 'mutants'. Back in the original box, the remaining coins (the E. coli cells) will continue to be tossed and, whenever tails turns up, the coin will be scooped out and transferred to the mutant box. Gradually, the mutant box will accumulate more and more coins. Translated back into the experiment, the mutants able to grow on lactose will continuously appear in the experiment, precisely as Cairns discovered.

We published our model in 1999, but it did not gather many converts. Undeterred, Johnjoe went on to write the book *Quantum Evolution*,[13] claiming a wider role for quantum mechanics in biology and evolution. But remember, this was before the role of proton tunnelling in enzymes was widely accepted, and quantum coherence hadn't yet been discovered in photosynthesis, so scientists were rightly sceptical about the idea of weird quantum phenomena being involved in mutation; and in truth, we skipped over several issues.[14] Also, the phenomenon of adaptive mutations became *messy*. It was discovered that the starved E. coli cells in Cairns's experiment were eking out a living from the trace nutrients of dead and dying cells and would occasionally replicate and even exchange DNA. Conventional explanations of adaptive mutations started to appear, which claimed to account for the raised mutation rates by a combination of several processes: a general increase in the mutation rate of all genes; cell death and release of the dead cells' mutated DNA; and, finally, selective uptake and amplification of the mutated lactose gene by surviving cells that managed to incorporate it into their genome.[15]

Whether these 'conventional' explanations can fully account for adaptive mutations remains unclear. Twenty-five years since Cairns's original paper appeared, the phenomenon remains puzzling, as evidenced by the continued appearance of papers investigating its mechanism,[16] not only in E. coli but also in several other microbes. As things currently stand, we don't exclude the possibility of quantum tunnelling being involved in adaptive mutations; but, at this time, we cannot claim that it is the only explanation.

In the absence of a strong need to implicate quantum mechanics in adaptive mutations, we recently decided to take a step back and investigate the more fundamental question of whether quantum tunnelling plays a role in mutation at all. As you will remember, the case for quantum tunnelling being involved in mutation was first made on theoretical grounds by Löwdin and has since been supported by several theoretical studies,[17] and also by experimental studies of what are called 'model base pairs', which are chemicals designed to have the same base-pairing properties as the bases in DNA, but are more amenable to experimentation. However, no one has yet proved that proton tunnelling causes mutation. The problem is that it has to compete with several other causes of mutations and mutation repair mechanisms, which makes unravelling its role, if it exists, all the more difficult.

To investigate this issue, Johnjoe has borrowed ideas from the enzyme experiments described in chapter 3, where you may recall the involvement of proton tunnelling was inferred after discovering 'kinetic isotope effects'. If quantum tunnelling is involved in speeding up an enzyme reaction, then replacing a hydrogen nucleus (a single proton) with a deuterium nucleus (consisting of a proton and a neutron) should slow the reaction since quantum tunnelling will be highly sensitive to doubling the mass of the particle trying to tunnel. Johnjoe is currently attempting a similar approach for mutation, investigating whether rates of mutation are different in deuterated water: D_2O, rather than H_2O. As we write, it seems the rates are indeed changed by the substitution; but much more work needs to be done to be sure that the effect is indeed due to quantum tunnelling, as replacing hydrogen with deuterium could affect many other bio-molecular processes without recourse to any quantum mechanical explanation.

Jim has focused on investigating whether quantum tunnelling of protons in the DNA double helix is feasible on theoretical grounds. When a theoretical physicist tackles a complex problem such as this, he or she tries to create a simplified model that is mathematically tractable while still retaining what are thought to be the most important features of the system or process. Such models can then be ramped up in sophistication and complexity as more of the details are added in order to get ever closer to mimicking the real thing.

The model chosen as the starting point for mathematical analysis in this case can be pictured as a ball (representing the proton) held in place by two springs attached to walls (figure 7.4), one on each side, pulling the ball in opposite directions. The ball tends to rest at the position where the pull from both springs is the same; so if one spring were slightly stiffer (less stretchy) than the other, then the ball would sit closer to the wall to which the stiffer spring is attached. However, there would still have to be some 'give' in this spring, such that it would also be possible for the ball to settle in a less stable position closer to the other wall. This then corresponds to what in quantum physics is called a *double potential energy well* and maps to the situation of a coding proton in the DNA strand, with the left-hand well in the diagram corresponding to the normal position of the proton, whereas the right-hand well corresponds to the rarer tautomeric position. Considered classically, although the proton will be found mostly in the left-hand well, if it receives an energetic

Figure 7.4: The proton of a hydrogen bond linking two DNA base sites can be regarded as being on two springs, such that it can oscillate from side to side. It has two possible stable positions, modelled here as a double energy well. The left-hand well (corresponding to the un-mutated position) is slightly deeper than the right-hand well (the tautomeric position), and so the proton prefers to sit in the left one.

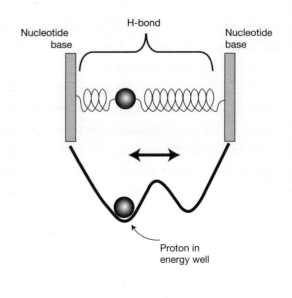

enough kick from an outside source, it can occasionally be knocked over to the other (tautomeric) side. But it will always be found in one well or the other. However, quantum mechanics allows the proton to spontaneously tunnel through the barrier, even if it has insufficient energy to clamber over the top: it doesn't necessarily need a kick. Not only that, but the proton can therefore be in a superposition of two position states (left and right wells) simultaneously.

Of course, drawing a picture is much easier than writing down a mathematical model that accurately describes the situation. To understand the proton's behaviour we need to map the shape of this potential well, or energy surface, very accurately. This is no trivial matter, as its precise shape depends on many variables. Not only is the hydrogen bond typically part of a large and complex DNA structure consisting of hundreds or even thousands of atoms, it is also immersed in a warm bath of water molecules and other chemicals inside the cell. Moreover, molecular vibrations, thermal fluctuations, chemical reactions initiated by enzymes and even ultraviolet or ionizing radiation can all affect the behaviour of the DNA bond both directly and indirectly.

One approach to tackling this level of complexity adopted by Adam Godbeer, a PhD student of Jim's, involves using a powerful mathematical approach, currently popular with physicists and chemists for modelling complex structures, called *density functional theory* (DFT). This allows the shape of the hydrogen bond's energy well to be calculated very accurately by taking into account as much of the structural information of the DNA base pair as is computationally possible. Think of DFT's job as providing a map of all the forces acting on the hydrogen bond due to the pulling, pushing and wobbling of the surrounding atoms of the DNA. This information is then used to calculate the way proton tunnelling behaves over time. An added complication is that the presence of the surrounding atoms in the DNA, and of water molecules, is continuously affecting the proton's behaviour and ability to quantum tunnel across from one strand of DNA to the other. But this constant influence of the external environment can also be included in the quantum mechanical equations. At the time of writing, in the summer of 2014, Adam's preliminary results suggest that, although it is possible for the two protons to tunnel across to their tautomeric positions in the A–T bond, the probability of their doing so is rather small. What the theoretical models do show, however, is that the action of the surrounding environment within the cell actively assists, rather than hinders, the tunnelling process.

What, then, can we surmise at the present moment about the link between quantum mechanics and genetics? We have seen that it is fundamental to heredity, since our genetic code is written in quantum particles. Just as Erwin Schrödinger predicted, quantum genes encode the classical structure and function of every microbe, plant and animal that has ever lived. This is not an accident, nor is it irrelevant, because high-fidelity copying of genes simply would not work if they were classical structures: they are too small not to be influenced by quantum rules. The quantum nature of genes allowed those Vostok microbes to faithfully replicate their genome over thousands of years, just as it allowed our ancestors to copy their genes over the many millions, indeed billions, of years that stretch back to the dawn of life on our planet. Life could not have survived and evolved on earth if it hadn't, billions of years ago, 'discovered' the trick of encoding information in the quantum realm.* On the other hand, whether quantum mechanics plays an important and direct role in genetic mutations – that *infidelity* in the copying of genetic information that is so vital for evolution – remains to be seen.

---

*This is currently a hot issue in quantum biology – namely, did life *discover* its quantum advantages, or is quantum mechanics just along for the ride?

8

Mind

JEAN-MARIE CHAUVET was born in the ancient French province of Auvergne, but when he was five his parents moved south-east to the Ardèche, a spectacular region of rivers, gorges and canyons cut into the underlying limestone rock. At twelve, Jean-Marie discovered his lifelong passion as he and his friends first donned Second World War helmets to explore the many caverns and caves dug out of the Ardèche valley walls along its great river. He left school at the age of fourteen, working first as a stonemason, then as a hardware store clerk and finally as a caretaker. Yet, inspired by Norbert Casteret's book *My Life Underground*, Jean-Marie devoted every weekend he could to his childhood passion, climbing across sheer rock faces or digging through dark caverns, dreaming of one day being the first person to set eyes on the hidden treasure of an unexplored cave. 'It's always the unknown that leads us. When you're walking along in a cave, you don't know what you're going to find. Will it end around the next corner, or will you discover something fantastic?'[1]

Saturday, 18 December 1994 started off a weekend like any other for the 42-year-old Jean-Marie and his two spelunking friends, Eliette Brunel-Deschamps and Christian Hillaire, roaming the gorges looking for something new. As the afternoon waned and the air grew colder they decided to explore an area known as the Cirque d'Estre, which captures the meagre afternoon sunlight and so is usually a little warmer than sheltered parts of the valley on a cold winter's day. The friends followed an old mule track that wound along the cliff between terraces of evergreen oak, box trees and heather, with a great view of the Pont d'Arc at the entrance of the gorge. As they struggled through the undergrowth they noticed a small cavity in the rock, measuring about 25 centimetres wide and 75 high.

This was – literally – an open invitation to the cavers, and they had soon squeezed through the gap to enter a small chamber only a few metres long and barely high enough to allow them to stand upright. Almost immediately, they noticed a faint draught coming from the back of the chamber. Anyone who has explored caves will be familiar with the sensation of a warm draught coming from an unseen tunnel. Most hidden passages are well known to experienced cavers; it's just that they lie beyond the narrow pencil of illumination provided by your torch. But the draught in that tiny chamber was not coming from any known cave. The team took turns removing stones from the end of the chamber until they located the source of the air: a duct falling vertically downward. The smallest of the team, Eliette, was the first to be lowered by rope into the darkness to a narrow shaft that she could crawl through. It first went down, then turned back up again before opening out, at which point Eliette could see that she was hanging 10 metres above a clay floor. Her torch was too weak to illuminate the far wall, but the echo that returned her shout from the darkness told her that she was in a big cave.

The team were very excited, but had to return to their van parked at the foot of the cliff to fetch a ladder. After retracing their steps back to the cavity they unrolled the ladder and Jean-Marie was the first to reach the floor of the cave. It was indeed a big cavern, at least 50 metres high and just as wide, with stunning columns of white calcite. The three carefully made their way through the darkness, stepping in each other's footprints to avoid disturbing the pristine environment, past great flows and curtains of mother of pearl and between the bones and teeth of long-dead bears scattered in ancient hibernation nests dug into the clay floor.

When the light of Eliette's torch reached the wall she let out a cry. She had spotted a line of red ochre forming the outline of a small mammoth. Speechless, the friends made their way along the wall, illuminating in turn the shapes of a bear, a lion, birds of prey, another mammoth, even a rhino and stencilled human hands. 'I kept thinking, "We're dreaming. We're dreaming,"' Chauvet remembered.[2]

The team's torches were losing power so they retraced their steps, crawled out of the cave and drove back to Eliette's home to have dinner with her daughter, Carole. But their emotional, disjointed and largely incoherent accounts of what they had seen so intrigued Carole that she insisted they take her straight back to the cave so she could see the marvels for herself.

It was after dark by the time they re-entered the cave, this time with more powerful torches which revealed the full splendour of their discovery: several caverns decorated with a marvellous menagerie of animals: horses, ducks, an owl, lions, hyenas, panthers, stags, mammoths, ibexes and bison. Most were drawn in a wonderful naturalistic style, with charcoal shading and overlapping heads to suggest perspective, and in poses that possess a real emotional appeal. There was a row of calmly pensive horses, a cute baby mammoth with large round feet and a pair of charging rhinos. There is even a rhino whose seven legs suggest a running motion.

The Chauvet cave, as it is now called, is today recognized as one of the world's most important sites of prehistoric art. Because it is so pristine – it even has the intact footprints of its ancient inhabitants – it remains sealed and guarded so as to preserve its delicate environment. Access is strictly controlled, with only a lucky few allowed to enter the cave: one of these was the German film-maker Werner Herzog, whose 2011 film, *Cave of Forgotten Dreams*, is the closest most of us will get to enjoying the remarkable rock art of the ice-age hunters who sheltered in those caves thirty thousand years ago.

What we wish to explore in this chapter is not the rock images themselves, but the puzzle probably best evoked by the title of Herzog's film. It is clear from any viewing of the paintings that they are not simply flat representations of what was seen by the eye. They are often abstracted to conjure an impression of motion, and they utilize bends and curves in the rock to endow the represented animals with an almost three-dimensional presence.* The artist(s) had not simply painted objects; they painted ideas. The humans who smeared pigment over the walls of the Chauvet cave were, like us, people who thought about the world and their place in it; they were conscious.

But what is consciousness? This is, of course, a question that has vexed philosophers, artists, neurobiologists and indeed the rest of us for probably as long as we have *been* conscious. In this chapter we will take the coward's way out by not attempting any rigid definition. Indeed, it is our view that the quest to understand this strangest of biological phenomena is often hindered by a pernickety insistence on defining it. Biologists cannot even agree on a unique definition of life itself; but that hasn't stopped them from unravelling aspects of the cell, the double helix, photosynthesis, enzymes and a host of other living phenomena, including many driven by quantum mechanics, that have now revealed a great deal about what it means to be alive.

We have explored many of these revelations in earlier chapters, but all those we have so far discussed, from magnetic compasses to enzyme action, from photosynthesis to heredity to olfaction, can be discussed in terms of conventional chemistry and physics. While quantum mechanics may be unfamiliar, particularly from many biologists' perspectives, it nevertheless fits completely within the framework of modern science. And although we may not have an intuitive or commonsense grasp of what is going on in the two-slit experiment or quantum entanglement, the mathematics that underpins quantum mechanics is precise, logical and incredibly powerful.

But consciousness is different. Nobody knows where or how it fits in with the kind of science that we have discussed so far. There are no (reputable) mathematical equations that include the term 'consciousness', and unlike, say, catalysis or energy transport, it has not,

---

* Shocking to many film buffs, Herzog's film is in 3D.

so far, been discovered in anything that isn't alive. Is it a property of *all* life? Most people would think not, and would reserve consciousness for those creatures that possess nervous systems; but then how much of a nervous system is necessary? Do clownfish yearn for their home reef? Did our European robin really feel an urge to fly south for the winter; or was she on automatic pilot like a drone aircraft? Most pet owners are convinced that their dogs, cats or horses are conscious; so did consciousness emerge in mammals? Many people who keep budgerigars or canaries are equally sure that their pets also have their own personalities and are just as conscious as the cats that chase them. But if consciousness is common to both birds and mammals, then both probably inherited the property from a common conscious ancestor, perhaps something like the primitive reptile called an *amniote* that lived more than three hundred million years ago and appears to be the ancestor of birds, mammals and dinosaurs. So, did the Tyrannosaurus rex that we met in chapter 3 experience fear as it sank into the Triassic swamp? And are more primitive animals really unconscious? Many aquarium owners would insist that fish or molluscs such as octopi are conscious; but to find an ancestor to all these groups we have to go back to the emergence of vertebrates in the Cambrian period five hundred million years ago. Is consciousness really that ancient?

Of course, we don't know. Even pet owners are only making guesses, since nobody really knows how to distinguish human-like behaviour from true consciousness. Without knowing what consciousness *is*, we can never know which life forms possess the property. So our naïve approach will be to eschew these arguments and debates and remain entirely agnostic on questions of when consciousness emerged on our planet, or which of our relatives in the animal kingdom are self-aware. We take as our starting point an insistence that those of our ancestors who painted the ideas of bears, bison or wild horses on ancient cave walls were definitely conscious. So, some time between three billion or so years ago, when microbes first emerged from the primeval mud, and the tens of thousands of years ago when those early modern humans decorated caves with impressions of animals, a bizarre property appeared in the matter of which living organisms are composed: some of that matter became aware. Our aim in this chapter will be to consider how and why this happened; and to examine the controversial suggestion that quantum mechanics played a key role in the emergence of consciousness.

First, in the spirit of our previous chapters, we will ask the question whether we *need* to resort to quantum mechanics in order to explain this most mysterious of human phenomena. It is certainly not enough to adopt the view, as some have, that consciousness is mysterious and difficult to pin down, and quantum mechanics is mysterious and difficult to pin down, therefore surely the two must be connected in some way.

How odd is consciousness?

Perhaps the oddest fact we know about the universe is that we know a great deal about it, owing to an extraordinary property possessed by those parts of it that are enclosed within our own skulls: our conscious minds. This is indeed highly bizarre, not least because the function of this odd property isn't at all clear.

Philosophers often probe this question by imagining the existence of zombies. These function just as human beings going about their activities, painting the walls of caves or reading books, but without any inner life; nothing is going on inside their heads except mechanical calculations that drive the movement of their limbs or the motor functions that power their language. Zombies are automatons, without awareness or sense of experience. That such beings are at least a theoretical possibility is evidenced by the fact that a lot of our actions – walking, riding a bicycle, the movements required to play a familiar musical instrument etc. – may be performed unconsciously (in the sense that our conscious mind can be elsewhere when performing these tasks), without awareness or recollection of experience. Indeed, when we actually think about them, our performance in these activities is paradoxically hampered. For these actions at least, consciousness seems to be dispensable. But if there exist activities that can be executed without consciousness, then is it at least possible to imagine a creature performing all human activities on automatic pilot?

It would seem not; there are some activities for which consciousness appears to be indispensable, such as natural language. It is very hard to imagine holding a conversation on automatic pilot. It would also be difficult for us to do a tricky calculation automatically or solve a crossword. We cannot imagine our ice-age artist (we will arbitrarily assume our painter was female) being able to paint a bison with nothing but the wall of a cave in front of her, if she were not conscious. What all of these necessarily conscious activities have in common is that they are driven by *ideas*, such as the idea behind a word, the solution to a problem, or an understanding of what a bison is and means to stone age people. Indeed, the walls of the Chauvet cave provide lots of evidence for that most powerful application of ideas: bringing several of them together to form a novel concept. A hanging rock, for example, is painted with an impossible figure that possesses a bison's upper body but a human lower half. Such an object could only have been created in a conscious mind.

What, then, are ideas? For our purposes, we will assume that ideas represent complex information that is joined up in our conscious mind to form concepts that have meaning to us, such as whatever it was that the half-man half-bison image on the wall of the Chauvet cave meant to the people who inhabited those caves. This compression of complex information into a singular idea was noted in a description attributed to Mozart of how an entire musical composition might be 'finished in my head though it may be long. Then my mind seizes it as a glance of my eye . . . It does not come successively, with various parts worked out in detail, as they will be later on, but in its entirety.'[3] The conscious mind is able to 'seize' on complex information 'with various parts' so that its meaning can be grasped 'in its entirety'. Consciousness allows our mind to be driven by ideas and concepts, rather than mere stimuli.

But how does complex neuronal information get glued together in our conscious minds to form an idea? This question is an aspect of the first puzzle of consciousness – what is often termed the *binding problem*: how does information encoded in disparate regions of our brain come together in our conscious mind? The binding problem is usually formulated in terms of visual or other sensory information. Recall, for example, Luca Turin's evocative description of the scent of Shiseido's Nombre Noir perfume: 'It was halfway between a rose and a violet, but without a trace of the sweetness of either, set instead against an austere, almost saintly background of cigar-box cedar notes.' Turin did not experience the perfume as a mix of distinct smells each associated with the firing of its particular olfactory receptor neuron, but as a single aroma with a range of underlying evocative notes and tones, including the meaning of a whole host of accessory concepts, such as cigars and violets. Similarly, sights and sounds are not experienced as distinct proportions of colours, textures or notes but as integrated sensory impressions, memories and concepts of, for example, a bison, a tree or a person.

Imagine our Palaeolithic artist observing a real bison. Her eyes, nose, ears and, if it was a dead bison, touch receptors in her fingers will have captured a multitude of sensory impressions of the animal, including its smell, shape, colour, texture, motion and sound. In chapter 5 we discussed how scents are captured by our sense of smell. You may recall that odorant molecules that bind to each olfactory neuron cause the nerve to 'fire', which means that it sends an electrical signal along its axon (the broom-handle end of the cell) which goes from the olfactory epithelium in the back of the nose to the olfactory bulb in the brain. We will be exploring the details of this firing process later in the chapter, because it is key to understanding the potential involvement of quantum mechanics in our thoughts. For now, however, we will imagine a bovine odorant molecule wafting off our bison and into the nose of our artist, where it binds to an olfactory receptor and triggers a chain of electrical pulses to travel down the wire-like axon, a bit like a telegraph signal but comprising only dots, or blips, rather than the dots and dashes of telegrams.

Once the olfactory nerve signal arrived in our artist's brain, it triggered the firing (more blips) of many more downstream nerves: the blip signal hopped from one nerve to another, with each nerve acting like a kind of telegraph relay station. Other sensory data were similarly captured into blip signals. For example, rods and cones (specialist neurons like olfactory neurons, but responding to light rather than odorants) lining the retina of her eye would have sent trains of blip signals via optical nerves to the visual cortex of her brain. And just as olfactory neurons responded to individual odorant molecules, optical nerves responded to only certain features of an image that fell on her retina: some will have responded to a particular colour or shade of grey, others to edges or lines or particular textures. Auditory nerves in her inner ear similarly responded to sound, perhaps the heavy breathing of a speared bison; and the touch of its fur would have been captured by *mechanosensitive* nerves in her skin. In all these instances, each sensory neuron would have responded only to certain features of the sensory input. For example, a particular auditory neuron would have fired only if the sound entering the artist's ears included a certain frequency. But, whatever its source, the signal generated by each nerve would have been precisely the same: a pulse of electrical blips travelling from the sensory organ to specialist regions of the artist's brain. There, those signals may have triggered immediate motor outputs; but they may also have modified the connectivity between neurons to lay down a memory of her observations via the 'neurons that fire together wire together' principle that appears to underpin how memories are encoded in the brain.

The important point is that there is nowhere in the 100 billion or so neurons of a human brain where this vast sensory stream of blips comes together to form the conscious impression of a bison. In fact, 'stream' isn't really the right word here, because it suggests some pooling of information within that stream, and that doesn't happen in neurons. Instead, each nerve signal remains locked into an individual nerve. So, rather than a stream, you should think of the information travelling through the brain as sequences of blip blip blip blip . . . signals passing along individual strands of an immense tangle of trillions of neurons. The binding problem is the problem of understanding how all the disparate blip-encoded information generates the unified perception of a bison.

And it isn't just sensory impressions that need to be bound. The raw material of consciousness is not sensory data stripped of context, but meaningful concepts – in the case of a bison, hairy, smelly, scary or magnificent – each of which is loaded with lots of complex information. All this additional baggage must have been bound together with the sensory impressions to provide the impression of a hairy, smelly, scary but magnificent bison that our Palaeolithic artist could have later recalled when reproducing it in pigment smears.[4]

Formulating the binding problem in terms of ideas, rather than sensory impressions, brings us to the nub of the problem of consciousness, which is the puzzle of how ideas can move minds and thereby bodies. We will never know what precisely was in the mind of our stone age artist that provoked her into applying pigment to stone. Perhaps she thought that the form of a bison would cheer up a dark corner; or maybe she believed that painting the animal would improve her fellow hunters' chances of success. But what we can be sure of is that the artist would have *believed* that the decision to paint the bison was her *idea*.

But how can an idea move matter? Understood as an entirely classical object, the brain receives information via one of the sensory inputs and then processes that information to generate outputs, just like a computer (or a zombie). But where in that tangle of blips is our conscious mind, that sense of 'self' that, we are convinced, drives our voluntary actions? What exactly is this *consciousness* and how does it interact with the matter of our brain to move our arms, legs or tongue? Consciousness, or free will, just doesn't figure in an entirely deterministic universe, because the laws of causality allow only for one thing after another in an endless chain of cause and effect that stretches from that Chauvet cave right back to the Big Bang.

Jean-Marie describes the moment when he and his friends first laid eyes on the paintings in the Chauvet cave: 'We were weighed down by the feeling that we were not alone; the artists' souls and spirits surrounded us. We thought we could feel their presence.'[5] Clearly the cavers were experiencing a profound, what some would call a spiritual, experience. When we look inside the skull of a human or animal all we find is wet squishy tissue, not very different from the stuff of a bison steak. But when that stuff is inside our own skull, it is aware and has experiences, concepts that just don't seem to exist in the material world. And somehow this ethereal stuff of awareness and experience – our conscious mind – drives the material stuff of our brain to *cause* our actions (or at least, that's our impression). This puzzle, variously referred to as the *mind–body problem* or the *hard problem* of consciousness, is surely the deepest mystery of our entire existence.

In this chapter we will be asking whether quantum mechanics can provide any answers to this deep mystery. We should emphasize at the outset that any ideas about consciousness remain highly speculative in nature, since no one really knows what it is or how it works. There isn't even a consensus among neuroscientists, psychologists, computer scientists and artificial intelligence researchers that there is a need for something beyond the sheer complexity of the human brain to explain consciousness.

Our starting point will be the brain processes that led to the shape of a bison being impressed on Ardèche limestone.

The mechanics of thought

In this section we will follow the causal chain backwards from the appearance of a line of red ochre on the wall of a cave thirty thousand years ago. This pursuit will lead us from the contracting muscles in the arm of the painter who drew that line, back to the nerve impulses that caused the muscles to contract, further back to the brain impulses that fired those nerves and the sensory inputs that set the chain of events in motion. Our aim is to try to pin down where consciousness makes its input in this causal chain, so that we can then investigate whether quantum mechanics might have played a role in that event.

We can imagine the scene all those millennia ago when an unknown artist, dressed perhaps in bearskins, peered into the gloom of the Chauvet cave. The paintings were discovered deep within the cave, so she would have had to carry a torch, along with pots of pigments, into the cave. Then at some point the painter dipped a finger into the pot of coloured charcoal and smeared the pigment onto the wall to create the outline of a bison.

The motion of the painter's arm across the cave wall was initiated by a muscle protein called myosin. Myosin is an enzyme that uses chemical energy to power the contraction of muscles, essentially by causing the fibres to slide over each other. The details of this contraction mechanism have been worked out by hundreds of scientists over several decades, and it is a remarkable example of nanoscale biological engineering and dynamics. But in this chapter we will skip the fascinating molecular details of muscle contraction to focus instead on the question of how something as ephemeral as an idea could cause muscles to contract (figure 8.1).

The immediate answer is that it didn't. The contraction of the artist's muscle fibres was actually triggered when positively charged sodium ions rushed into her muscle cells. Muscle cells have more sodium ions on the outside of their membrane than the inside, giving rise to a voltage difference across their membrane, a bit like a tiny

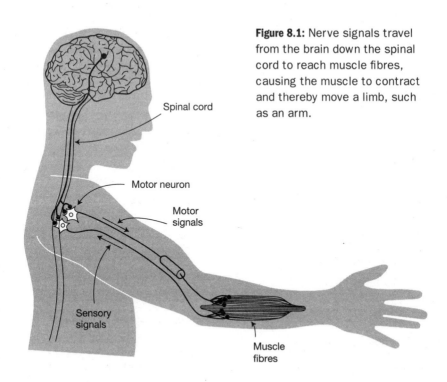

Figure 8.1: Nerve signals travel from the brain down the spinal cord to reach muscle fibres, causing the muscle to contract and thereby move a limb, such as an arm.

Spinal cord

Motor neuron

Motor signals

Sensory signals

Muscle fibres

battery. However, there are pores in these membranes called *ion channels*, which, if opened, allow the sodium ions into the cell. It was this electrical discharging process that triggered the artist's muscle contraction.

The next backward step in our chain of causation is the question: what caused those muscle ion channels to flip open at that moment? The answer is that *motor nerves* attached to the muscles in the artist's arm released chemicals called neurotransmitters that popped the ion channels open. But what then caused these motor nerves to release their package of neurotransmitters? Nerve endings release neurotransmitters whenever an electric signal called an *action potential* arrives (figure 8.2). Action potentials are fundamental to all nerve signalling, so we need to take a closer look at how they work.

A nerve cell, or *neuron*, is an extremely long, thin, snake-like cell consisting of three parts. At its head end is a spider-like *cell body*, which is where the action potential is initiated. This then travels

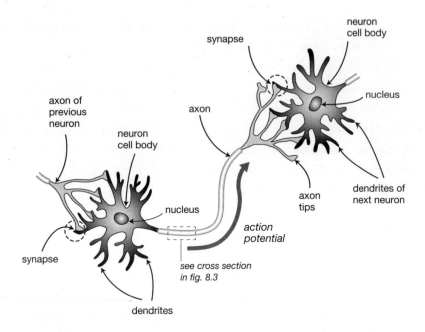

Figure 8.2: Nerves send electrical signals from the cell body along the axon to the nerve ending, where they cause the release of neurotransmitter into a synapse. The neurotransmitter is picked up by the cell body of a downstream neuron, causing it to fire and thereby transmit the nerve signal from one neuron to the next.

along the thin middle section, called the *axon* (the 'broom handle' of an olfactory neuron), to the nerve ending, where the neurotransmitter molecules are released (figure 8.2). Although the nerve axon looks a little like a tiny electric cable, the way it transmits its electrical signal is far cleverer than the process by which a simple flow of negatively charged electrons passes through a copper wire.

The nerve cell, just like a muscle cell, normally has more positively charged sodium ions outside than inside. This difference is maintained by pumps that push positively charged sodium ions out of the cell through the nerve cell membrane. The excess of external positive charges provides a voltage difference across the cell membrane of about one-hundredth of a volt. Although this doesn't sound like much, you have to remember that cell membranes are just a few nanometres thick, so it is a voltage across a very short distance. This means that we have an electrical gradient (what voltage actually is) across the cell membrane of a million volts per metre. This is equivalent to a staggering ten thousand volts across a one-centimetre gap and is almost enough to create a spark, such as is required in your car's spark plug to ignite the fuel.

The head-end of the artist's motor nerve, the body of the nerve cell, is connected to a cluster of structures called *synapses* (figure 8.2), which are kind of nerve-to-nerve junction boxes. Upstream nerves release neurotransmitter molecules into these junctions much as neurotransmitters are released at the nerve–muscle junction; this triggers the opening of ion channels in the membrane surrounding the nerve cell body membrane, thereby allowing positively charged ions to rush inside, causing its voltage to drop sharply.

Most voltage drops caused by the opening of a handful of ion channels in a synapse will have little or no effect. But if lots of neurotransmitter arrives, then lots of ion channels will flip open. The ensuing rush of positive ions into the cell causes its membrane voltage to dip below a critical threshold of about −0.04 volts. When this happens, another set of nerve ion channels come into play. These are *voltage-gated ion channels*, which means they are sensitive, not to neurotransmitters, but to the voltage difference across the membrane. In the example of our artist, when the voltage in the cell body dropped below the critical threshold, a whole bunch of these channels opened to allow more ions to rush into the nerve, further short-circuiting their patch of membrane. The ensuing voltage drop caused more voltage-gated ion channels to pop open, allowing more ions to rush inside the cell, causing more of the membrane to short-circuit. The long cable of the nerve, the axon, is lined with these voltage-gated channels, so once the short-circuiting was kicked off at the cell body, it triggered a kind of domino effect of membrane short-circuiting – the action potential – that quickly travelled down the nerve until it reached the nerve ending (figure 8.3). There it stimulated the release of neurotransmitter into the *neuromuscular*

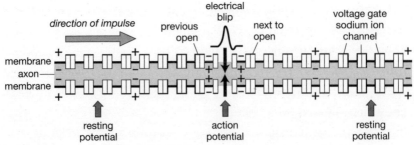

Figure 8.3: Action potentials travel along nerve axons via the action of voltage-gated ion channels in the nerve cell membranes. In its resting state the membrane has more positive ions on the outside than on the inside. However, a change in voltage caused by an upstream action potential will trigger opening of the ion channels and a surge of positively charged sodium ions – an action potential – will rush into the cell, temporarily reversing the membrane voltage. This electrical blip will trigger the opening of downstream ion channels in a kind of domino effect electrical impulse that travels down the nerve until it reaches the nerve ending, where it triggers neurotransmitter release. After the action potential has passed, ion pumps return the membrane to its normal resting state.

junction, causing our artist's arm muscle to contract to trace the line of a bison on the wall of the cave (figure 8.1).

You can see from this description how different nerve signals are from an electrical signal travelling down a wire. For a start, the current, the movement of charges, is not down the length of nerve cables in the direction of the nerve signal, but perpendicular to the direction of the action potential: from outside in, through those ion channels in the cell membrane. Also, immediately after the action potential is initiated by the opening of the first ion channels, they are slammed shut again and the ion pumps get to work on re-establishing the original battery voltage across the membrane. So another way of viewing the nerve signal is as a wave of opening and closing of membrane ion doors that travels from the cell body to the nerve ending: a moving electrical blip.

The nerve–nerve junctions for most motor nerve cells are located in the spinal cord, where they receive neurotransmitter signals from hundreds or even thousands of upstream nerves (figure 8.1). Some upstream nerves release neurotransmitters into the junction box (synapse) that open ion channels in the cell body to increase the likelihood of firing up the motor nerve, whereas others tend to close them. In this way the cell body of each nerve cell seems to be acting like the logic gate of a computer, generating an output – whether or not it fires – based on its inputs. So, if the neuron is like a logic gate, then the brain, made up of billions of neurons, might be thought of as some kind of computer; or at least, this is the assumption of most cognitive neuroscientists who subscribe to what is called the *computational theory of mind*.

But we are jumping too far ahead – we haven't yet reached the brain. Our artist's motor nerve must have received lots of neurotransmitters in its nerve–nerve junction boxes, causing it to fire. Those inputs came from upstream nerves that mostly originated in her brain. Following the chain of causation back, the heads of those nerves would have made their decisions about whether or not they fired on the basis of their many inputs, and the inputs of those inputs, and so on further and further backwards through the causal chain until we reach the nerves that received input signals from the artist's eyes, ears, nose and touch receptors, and memory centres that would have received sensory inputs from her earlier observations of live and dead bison. Between sensory inputs and motor output is the brain's *neural network* that performed the computations dictating the decision to generate, or not to generate, the precise motor output needed to draw the outline of a bison.

So there we have it: the entire chain of events leading up to that muscle contraction that swept the artist's arm across the wall. But have we missed something? What we have described so far is an entirely mechanistic causal chain from sensory input to motor output, with some of the information channelled through memory centres. This is the kind of mechanism that Descartes was talking about when he made the claim (discussed in chapter 2) that animals are mere machines; all we've done is replace his pulleys and levers with nerves, muscles and logic gates.

But remember that Descartes reserved a role for a spiritual entity, the soul, as the ultimate driver of human actions. Where is the soul in this input–output chain of events? So far, we have described only a zombie artist. Where did her consciousness, her idea that she should represent a meaningful bison on the wall of the cave, enter the chain of events between input and output? This remains the biggest puzzle of brain science.

How mind moves matter

In one way or another, most people probably subscribe to the notion of *dualism* – the belief that the mind/soul/consciousness is something other than the physical body. But dualism fell out of favour in scientific circles in the twentieth century, and most neurobiologists now prefer the idea of *monism* – the belief that mind and body are one and the same thing. For example, the neuroscientist Marcel Kinsbourne claims that 'being conscious is what it is like to have neural circuitry in particular interactive functional states'.[6] But the logic gates of a computer are, as we have already noted, rather similar to neurons, so it isn't clear why highly connected computers, such as the world-wide web with its one billion or so internet hosts (though still small compared to the brain's one hundred billion neurons), show no sign of awareness. Why are silicon-based computers zombies whereas flesh-based computers are conscious? Is it simply a matter of complexity and the sheer 'interconnectedness' of our brain cells, not yet matched by the world-wide web,* or is consciousness a very different kind of computing?

There are of course many *explanations* of consciousness, all of which have been laid out in a whole host of books on the topic.

---

* The size of the internet is not easy to estimate, but each web page currently links to, on average, fewer than a hundred other pages, whereas neurons have synaptic links to thousands of other neurons. So, in terms of links, there are about a trillion between web pages and about a hundred times that number between neurons in the human brain. But the web doubles in size every few years, so it is anticipated that it will rival the complexity of the human brain within a decade. Will the internet then become conscious?

But, for the purposes of this account, we will focus on the highly controversial, yet fascinating, claim that is most relevant to our theme: namely, that consciousness is a quantum mechanical phenomenon. The case was most famously made by the Oxford mathematician Roger Penrose who, in his 1989 book *The Emperor's New Mind*, claimed that the human mind is a quantum computer.

You may remember the idea of quantum computers from chapter 4, where we recalled that *New York Times* article of 2007 claiming that plants were quantum computers. The MIT team eventually came round to the idea that microbes and plant photosynthesis systems may indeed be performing some kind of quantum computation. But could their own very clever brains have also been operating in the quantum realm? To examine this question we first need to take a closer look at what quantum computers are, and how they work.

Computing with qubits

When we think of a computer today we mean any electronic device capable of carrying out instructions to manipulate and process information via a collection of electrical switches that can be either ON or OFF – each capable of encoding a binary digit (or *bit*) as a 1 or a 0. A collection of such switches can be arranged to build circuits that perform logic instructions, which can be combined and used to carry out arithmetical operations such as addition and subtraction or indeed the opening and closing of gates that we described for neurons. The great advantage of this electrical *digital computer* is that it is very much faster than any manual way of performing the same kind of task, whether by counting on fingers, mental arithmetic or using a pen and paper.

But while electronic computers may be extraordinarily fast at doing sums, even they cannot keep track of the complexity of the quantum world with its multitude of overlapping probabilities. To overcome this problem, the Nobel Prize-winning physicist Richard

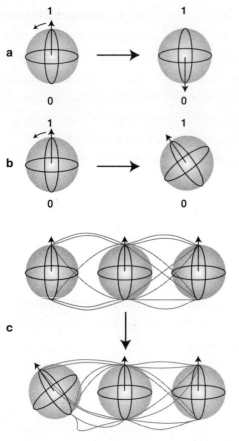

Figure 8.4: (a) A classical bit being switched from 1 to 0 is represented as a rotation of a classical sphere through 180°. (b) A qubit being switched may be represented as a rotation of a sphere through any arbitrary angle. However, a coherent qubit may also be in a superposition of many rotations. (c) Three coherent qubits showing their entanglement interactions as imaginary strings connecting the surface of each sphere. It is the tension on these strings following rotations which instantiates quantum calculations.

Feynman came up with a possible solution. He suggested performing calculations in the quantum world, with a quantum computer.

To see how quantum computers might work, it will be useful first to represent the 'bit' of a classical computer as a kind of spherical compass whose needle may point at either 1 (north pole) or 0 (south pole) and is capable of turning through 180° to switch between these two states (figure 8.4a). The central processing unit (CPU) of a computer consists of many millions of these one-bit switches, so the entire computational process can be envisaged as the application of a complex set of switching rules (algorithms) that can flip lots and lots of spheres by 180°.

The quantum computing equivalent of the bit is called a *qubit*.

This is similar to the classical sphere,* but its movement is not limited to a 180° flip. Instead, it can rotate through any arbitrary angle in space and, being quantum mechanical, it can also point in many directions simultaneously in a quantum coherent superposition (figure 8.4b). This increased flexibility allows a qubit to encode more information than a classical bit. But the real boost to computing power comes when you put qubits together.

Whereas the state of one classical bit has no influence on its neighbours, qubits may also be *quantum entangled*. You may remember from chapter 6 that entanglement is a quantum step up from coherence whereby quantum particles lose their individuality, so that what happens to one affects them all, instantaneously. From the perspective of quantum computing, entanglement can be visualized as each qubit sphere being connected by elastic strings† to every other qubit (figure 8.4c). Now, let us imagine that we rotate just one of the spheres. Without entanglement, the rotation will not affect neighbouring qubits. But if our qubit is entangled with other qubits, then the rotation changes the tensions in all the connecting strings between these connected qubits. The computational resource of all those entanglement strings increases *exponentially* with the number of qubits, which means that it increases very rapidly indeed.

To get a feeling for exponential growth, you may have heard the fable about the Chinese emperor who was so pleased by the invention of chess that he promised to reward its inventor with a prize of his own choosing. The canny inventor asked for just one grain of rice for the first square on the chessboard, two grains of rice for the second, four for the third and so on, doubling the number of grains with each successive square until he reached the sixty-fourth square. The emperor, considering this to be a modest request, eagerly agreed and ordered his servants to bring out the rice. But, when the rice grains were counted out, he soon discovered his error. The first row

---

* For the physicist reader, what we are describing here is a Bloch sphere.
† In reality the strings represent the mathematical relationship between the phase and amplitude of the entangled qubits instantiated in the Schrödinger equation.

of squares amassed only 128 grains (2^7 plus one – remember, the first square has only a single rice grain) and even by the end of the second row of squares he had to find only 32,768 grains, just less than a kilogram of rice. But as the kilograms begin to multiply on subsequent squares, the emperor was dismayed to discover that by the end of the third row he had to hand over more than two hundred tons of rice. Reaching even the end of the fourth row would have bankrupted the kingdom! In fact, to reach the end of the chessboard would have required 9,223,372,036,854,775,808 (2^{63} plus one) grains of rice, or 230,584,300,921 tons, which is roughly equivalent to the entire world's rice harvest throughout the history of humankind.

The problem for the emperor was his failure to realize that doubling a number again and again leads to exponential growth – which is another way of saying that the increase from one number to the next is proportional to the size of the previous number. Exponential growth is explosive growth, as the emperor discovered to his cost. And just as the rice grains in the fable increased exponentially with the number of chessboard squares, so the power of a quantum computer scales exponentially with its number of qubits.

This is very different from a classical computer, whose power increases only *linearly* with the number of bits. For example, adding one more bit to an 8-bit classical computer will increase its power by a factor of one-eighth; to double its power, the number of bits will have to be doubled. But simply adding one qubit to a quantum computer will double its power, leading to the same kind of exponential increase in power that the emperor saw running away with his rice grains. In fact, if a quantum computer could maintain coherence and entanglement within just 300 qubits, which could potentially involve just 300 atoms, it could outperform, on certain tasks, a classical computer the size of the entire universe!

But, and this is a very big but, for the quantum computer to work, the qubits must interact only with each other to perform calculations (via their invisible entangled 'strings'). This means they must be completely isolated from their environment. The problem is that any interaction with the outside world causes the qubits to become entangled with their environment, which we can envisage as the formation of many more strings, all pulling on the qubits from different directions, competing with the strings between the qubits and therefore interfering with the calculation they are performing. This, essentially, is the process of decoherence (figure 8.5). With even the faintest interaction, the environment throws such a confusion of entanglement strings over the qubits that they cease to behave in a coherent fashion with each other: their quantum strings are effectively severed and the qubits will behave as independent classical bits.

Quantum physicists do their best to maintain coherence in the entangled qubits by working with very rarefied and carefully controlled physical systems, encoding qubits in a handful of atoms, cooling the system to within a fraction of absolute zero and surrounding their apparatus with extensive lagging to shut out any environmental influence. Using these approaches they have delivered

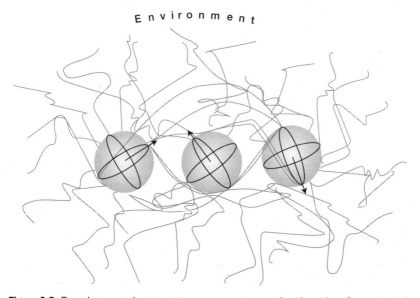

E n v i r o n m e n t

Figure 8.5: Decoherence in a quantum computer can be thought of as caused by entanglement of qubits with a tangle of environmental strings. These tug and pull at the qubits this way and that so they no longer respond to their own entanglement connections.

some landmark achievements. In 2001, scientists from IBM and Stanford University managed to build a seven-qubit 'test tube quantum computer' that could implement a clever code called Shor's Algorithm, named after the mathematician Peter Shor, who devised it in 1994 specifically to be run on a quantum computer. Shor's Algorithm encodes a very efficient way of factorizing numbers (working out what prime numbers need to be multiplied to give the required number). This was a huge breakthrough and made scientific headlines around the world; yet the maiden flight of this fledgling quantum computer managed only to compute the prime factors of the number 15 (3 and 5, in case you were wondering).

Over the past decade, some of the top physicists, mathematicians and engineers have worked hard to build bigger and better quantum computers, but progress has been modest. In 2011, Chinese researchers managed to factorize the number 143 (13 × 11), using just four qubits. Like the US group before them, the Chinese team used a system in which qubits were encoded in the spin states of atoms. A quite different approach has been pioneered by the Canadian company D-Wave, which encodes qubits in the motion of electrons in electrical circuits. In 2007, the company claimed to have developed the first commercial 16-qubit quantum computer, able to solve a Sudoku puzzle and other pattern-matching and optimization problems. In 2013, a collaboration of NASA, Google and the Universities Space Research Association (USRA) purchased (for an undisclosed sum) a 512-qubit machine built by D-Wave that NASA plans to use to search for exoplanets, that is, planets orbiting not our sun but distant stars. However, the problems so far tackled by the company have all been within the reach of conventional computer power, and many quantum computing experts remain unconvinced that D-Wave's technology is really quantum computing – or if it is, whether its design would ever make it any faster than a classical computer.

Whatever approach the experimenters choose to take, the challenges facing them in turning the current generation of fledgling quantum computers into something useful remain immense. The biggest problem is scaling up. Doubling qubits doubles quantum computation power, but it also doubles the difficulty of maintaining quantum coherence and entanglement. Atoms have to be colder, shielding has to be more effective, and it becomes more and more difficult to maintain coherence for more than a few trillionths of a second. Decoherence sets in well before the computer manages to complete even the simplest calculation. (Although at the time of writing the record for room temperature quantum coherence of nuclear spin states is an impressive 39 minutes.[7]) But, as we have discovered, living cells do manage to keep decoherence at bay for long enough to transport excitons in photosynthetic complexes, or electrons and protons in enzymes. Could decoherence similarly be kept at bay in the central nervous system to allow quantum computation to be performed in the brain?

Computing with microtubules?

Penrose's initial argument that the brain is a quantum computer came from a rather surprising direction: the famous (at least in mathematical circles) set of incompleteness theorems put forward by the Austrian mathematician Kurt Gödel. These theorems were very shocking to mathematicians in the 1930s who had confidently embarked on a programme to identify a powerful set of mathematical axioms that could prove true statements were true and false statements were false – basically, that the whole of arithmetic was internally consistent and free of any self-contradictions. It sounds like the sort of thing that only mathematicians or philosophers would worry about, but it was and continues to be a big deal in the field of logic. Gödel's incompleteness theorems showed that such an endeavour was doomed to failure.

The first of his theorems demonstrated that logical systems, such as natural language or mathematics, can make some true statements that they can't prove. This may seem an innocuous assertion, but its implications are very far-reaching. Consider a familiar logical system, such as language, which is capable of reasoning through statements such as 'All men are mortal. Socrates is a man' to conclude that

'Socrates is mortal'. It's easy to see, and easy to formally prove, that the last statement follows logically from the first two, given a simple set of algebraic rules (if A = B and B = C then A = C). But Gödel showed that any logical system complex enough to prove mathematical theorems has a fundamental limitation: application of their rules can generate statements that are true, but these statements cannot be proved with the same tools that were used to generate them in the first place.

This seems rather odd, and indeed it is. However, and this is important, Gödel's theorem does not mean that some true statements are simply not provable. Instead, one set of rules may be able to prove the truth of statements generated by, and therefore unprovable with, any other set of rules. For example, true but unprovable language statements may be provable within the rules of algebra; and vice versa.

This is, of course, a huge oversimplification that does not do justice to the subtleties of the subject. The interested reader might like to try the 1979 book on this and related subjects by the American professor of cognitive science Douglas Hofstadter.[8] The key point here is that in his book *The Emperor's New Mind*, Penrose takes Gödel's incompleteness theorems as the starting point for his argument, by first pointing out that classical computers use formal logical systems (computer algorithms) to make their statements. It follows from Gödel's theorem that they must also be capable of generating true statements they can't prove. But, Penrose argues, humans (or at least those members of the species who are mathematicians) can prove the truth of these unprovable but true computer statements. Therefore, he argues, the human mind is more than just a classical computer, since it is capable of what he calls non-computable processes. He then postulates that this non-computability requires something extra, something that can only be provided by quantum mechanics. Consciousness, he argues, requires a quantum computer.

This is, of course, a very bold claim to make on the grounds of the provability or not of a difficult mathematical statement, a point to which we will return. But in his later book *The Shadows of the Mind*, Penrose went even further to propose a physical mechanism by which the brain might calculate its sums in the quantum world.[9] He teamed up with Stuart Hameroff,* Professor of Anesthesiology and Psychology at the University of Arizona, to claim that structures called *microtubules* that are found in neurons are the qubits of quantum brains.[10]

Microtubules are long strings of a protein called tubulin. Hameroff and Penrose proposed that these tubulin proteins – the beads on the string – are capable of flipping between at least two different shapes, extended and contracted, and, crucially, are able to behave as quantum objects that exist in a superposition of both shapes at once to form something akin to qubits. Not only that, they postulated that tubulin proteins in one neuron are entangled with tubulin proteins in lots of other neurons. You will remember that entanglement is that 'spooky action at a distance' that potentially connects objects that are very far away from each other. If spooky connections between all the trillions of neurons in a human brain were possible, then they could, potentially, bind together all the information encoded in separated nerves and thereby solve the binding problem. They could also provide the conscious mind with the elusive but extraordinary powerful capabilities of a quantum computer.

There is much more to the Penrose–Hameroff consciousness theory, including, possibly even more controversially, a proposed involvement of gravity.† But is it credible? We, together with nearly all neurobiologists and quantum physicists, are far from convinced. One of the most obvious objections may be clear from the preceding description of how information travels from the brain through to the nerves. You may have noticed that we did not mention microtubules in that description. That is because it was unnecessary to do so, since they do not, as far as is known, have any direct role in neural

---

* Johnjoe would like to take the opportunity to apologize to Stuart Hameroff for spelling his name wrongly in his book *Quantum Evolution*.
† This is another difficult concept, but Penrose proposed an entirely idiosyncratic interpretation of the measurement problem in quantum mechanics by postulating that for sufficiently complex (and therefore more massive) quantum systems, their gravitational effect on space-time creates a disturbance that collapses the wave function, transforming quantum into classical systems, and that this process generates our thoughts. Details of this extraordinary theory are well described in Penrose's books, but it is fair to say that his proposal has, to date, few adherents in the quantum physics community.

information processing. Microtubules support the architecture of each neuron and transport neurotransmitters up and down its length; but they are not thought to be involved in the network-based information processing responsible for brain computations. So microtubules are unlikely substrates for our thoughts.

But perhaps an even more important objection is that brain microtubules are highly unlikely candidates as coherent quantum qubits simply because they are too big and complicated. In previous chapters we made a case for quantum coherence, entanglement and tunnelling in a whole range of biological systems from photosynthetic systems to enzymes, smell receptors, DNA and the elusive organ of magnetoreception in birds. But a key feature of all of these is that the 'quantum' part of the system (the exciton, electron, proton or free radical) is simple. It consists of either a single particle or small numbers of particles that do what they do over atomic-scale distances. This corresponds of course to Schrödinger's seventy-year-old insight that the kinds of living system that are likely to support quantum rules will involve small numbers of particles.

But the Penrose–Hameroff theory proposes that entire protein molecules composed of millions of particles are in quantum superposition and entangled not only with molecules within the same microtubule but with microtubules, similarly composed of millions of particles, in billions of nerve cells across the entire volume of the brain. This is very far from being plausible. Although no one has managed to measure coherence in brain microtubules, calculations suggest that quantum coherence in even single microtubules could not be maintained for timescales longer than a few picoseconds,[11] far too fleeting a time to have any impact on brain computation.*

However, perhaps an even more fundamental problem with the Penrose–Hameroff quantum consciousness theory is Penrose's original case for the brain being a quantum computer. You will remember that Penrose based this claim on his assertion that humans

---

* A picosecond is one millionth of one millionth (or 10^{-12}) of a second.

can prove Gödelian statements whereas computers can't. But this implicates quantum computation in the brain only if quantum computers can prove Gödelian statements better than a classical computer; not only is there absolutely no evidence for this assertion, but most researchers believe the contrary.[12]

A further point is that it is not at all clear that a human brain *can* actually perform any better than a classical computer in proving Gödelian statements. Although humans may be able to prove the truth of an unprovable Gödelian statement generated by a computer, it is equally possible that computers may be able to prove the truth of an unprovable Gödelian statement generated by a human mind. Gödel's theorem only limits the ability of one system of logic to prove all its own statements; it does not place limits on the ability of one system of logic to prove Gödelian statements generated by another.

But does that mean that there is no role for quantum mechanics in the brain? Is it likely that, with so much quantum action going on elsewhere in our bodies, our thoughts are driven entirely by the steam-engine processes of the classical world? Perhaps not. Recent research suggests that quantum mechanics may indeed play a crucial role in how the mind works.

Quantum ion channels?

A possible site for quantum mechanical phenomena in the brain lies within ion channels in neuronal cell membranes. As we have already described, these are responsible for mediating the action potentials – the nerve signals – that transmit information in the brain, so they play a central role in neural information processing. The channels are only about one-billionth of a metre long (1.2 nanometres) and less than half that wide, so the ions have to pass through them in single file. Yet they do so at an extraordinarily high rate of about a hundred million per second. And the channels are also highly select-ive. For example, the channel responsible for allowing potassium ions into the cell allows about one sodium ion through for every ten thousand potassium ions, despite the fact that the sodium ion is a little smaller than potassium – so you might naively expect it to easily slip though anything big enough to accommodate a potassium ion.

These very high transport rates, coupled with the extraordinary degree of selectivity exercised, underpins the speed of action potentials and, thereby, their ability to transmit our thoughts around our brain. But how ions are transported so rapidly and selectively has remained something of a mystery. Could quantum mechanics help? We have already discovered (in chapter 4) that quantum mechanics can enhance energy transport in photosynthesis. Can it also enhance ion transport in the brain? In 2012 the neuroscientist Gustav Bernroider, from the University of Salzburg, teamed up with Johann Summhammer from the Atom Institute at the Vienna University of Technology to perform a quantum mechanical simulation of an ion passing through a voltage-gated ion channel and discovered that the ion is delocalized (spread out) when it travels through the channel: more of a coherent wave than a particle. Also, this ion wave oscillates at very high frequencies and transfers energy to the surrounding protein by a kind of resonance process, so that the channel effectively acts as an *ion refrigerator* that reduces the kinetic energy of the ion by about half. This effective cooling of the ion helps to maintain its delocalized quantum state by keeping decoherence at bay and thereby promotes rapid quantum transport through the channel. It also contributes to selectivity, since the degree of refrigeration will be very different if potassium is replaced with sodium: constructive interference can promote potassium ion transport while destructive interference can inhibit sodium ion transport. The team concluded that quantum coherence plays an 'indispensable' role in the conduction of ions through nerve ion channels, and is thereby an essential part of our thinking process.[13]

We should emphasize that these researchers have not suggested that quantum coherent ions are capable of acting as any kind of neural qubits, nor have they suggested that they could play a role in consciousness; and, at first sight, it is hard to see how they could contribute to solving some of the problems of consciousness, such as the binding problem. However, unlike the microtubules in the Penrose–Hamerhoff hypothesis, the ion channels do at least play a clear role in neural computation – they underpin action potentials – so their state will reflect the state of the nerve cell: if the nerve is firing, then ions will be flowing (remember, they are moving as quantum waves) rapidly through the channels, whereas if the nerve is resting, any ions in the channels will be stationary. So, since the total sum of firing and non-firing neurons in our brain must somehow encode our thoughts, then those thoughts are also reflected – encoded – in the sum of all that quantum flow of ions into and out of nerve cells.

But how might the individual thought processes be combined to generate conscious, bound-up thoughts? One coherent ion channel – whether quantum or classical – can't possibly encode all the information bound into the thought processes that culminate in visualizing a complex object, such as a bison. To play a role in consciousness, ion channels would have to be linked in some way. Could quantum mechanics help? Is it possible, for example, that the ions in a channel are not only coherent along the length of the channel but also coherent or even entangled with ions in adjacent channels or even nearby nerve cells? Almost certainly not. Ion channels and the ions within them would suffer the same problem as the Penrose–Hameroff microtubule idea. Although it is just about conceivable that a single ion channel could be entangled with an adjacent channel within the same nerve cell, entanglement between ion channels in different nerves, which would be needed to solve the binding problem, is totally unfeasible in the warm, wet, highly dynamic and decoherence-inducing environment of a living brain.

So, if entanglement can't bind the quantum-level information in ion channels, is there anything else that could do the job? There may be. Voltage-gated ion channels are of course sensitive to voltage: it's what opens and closes the channels. Voltage is just a measure of the gradient of an electric field. But the entire volume of the brain is filled with its own electromagnetic (EM) field, which is generated by the electrical activity of all its nerves. This field is what is routinely detected by brain-scanning technologies such as electroencephalography (EEG) or magnetoencephalography (MEG), and even a glance at one of those scans will tell you just what an extraordinarily complex and information-rich field it is. Most neuroscientists have ignored the potential role that the EM field might play in brain computation because they have assumed it is like the steam whistle of a train: a product of brain activity, but with no impact on that activity. However, several scientists, including Johnjoe, have recently seized upon the idea that shifting consciousness from the discrete particles of matter in the brain to the joined-up EM field could potentially solve the binding problem and provide a seat for consciousness.[14]

To understand how this might work, we probably need to say a bit more about what we mean by a *field*. The term derives from its common usage: it means something that is extended through space, like a cornfield or a football field. In physics, the term 'field' has the same essential meaning, but usually refers to energy fields that are able to move objects. Gravitational fields move anything that has mass, and electric or magnetic fields move electrically charged or magnetic particles such as the ions in nerve ion channels. In the nineteenth century James Clerk Maxwell discovered that electricity and magnetism are two aspects of the same phenomenon, electromagnetism, so we refer to both as EM fields. Einstein's equation, $E = mc^2$, with energy on one side and mass on the other, famously demonstrated that energy and matter are interchangeable. So the brain's EM energy field – the left-hand side of Einstein's equation – is just as real as the matter that makes up its neurons; and, because it is generated by neuron firing, it encodes exactly the same information as the neural firing patterns of the brain. However, whereas neuronal information remains trapped in those blipping neurons, the electrical activity generated by all the blipping unifies all the information within the brain's EM field. This could potentially solve the binding problem.[15] And, by opening and closing the voltage-gated ion channels, the EM field couples to those quantum coherent ions travelling through the channels.

When EM field theories of consciousness were first proposed at the very beginning of the present century, there was no direct evidence that the brain's EM field could influence nerve firing patterns to drive our thoughts and actions. However, experiments carried out in several laboratories have recently demonstrated that external EM fields, of similar strength and structure to those that the brain itself generates, do indeed influence nerve firing.[16] In fact, what the field seems to do is to coordinate nerve firing: that is, bring lots of neurons into synchrony so that they all fire together. The findings suggest that the brain's own EM field, generated by nerve firing, also influences nerve firing, providing a kind of self-referencing loop that many theorists argue is an essential component of consciousness.[17]

Synchronization of nerve firing by the brain's EM field is also very significant in the context of the puzzle of consciousness because it is one of the very few features of nerve activity that is known to correlate with consciousness. For example, we have all experienced the phenomenon of looking for an object that is in plain sight, such as our glasses, and then spotting it among a jumble of other objects. While we were looking at that jumble, the visual information encoding that object was travelling through our brain, via our eyes, but somehow we didn't see the object we were searching for: we were not *conscious* of it. But then we do see it. What changes in our brain between the times when we are first unconscious and then conscious of an object within the same visual field? Remarkably, neural firing itself doesn't seem to change: the same neurons fire whether or not we *see* the glasses. But when we don't spot our glasses, the neurons fire asynchronously and when we do they fire synchronously.[18] The EM field, pulling together all those coherent ion channels in disparate parts of the brain to generate synchronous firing, could play a role in this transition between unconscious and conscious thoughts.

We should stress that invoking ideas such as brain EM fields, or indeed quantum coherent ion channels, in order to explain consciousness does not in any way provide support for so-called 'paranormal phenomena' such as telepathy, since both concepts are only capable of influencing neural processes going on *inside* a single brain – they do not allow communication between different brains! And, as we have pointed out when considering Penrose's Gödelian argument, there is in fact no evidence that quantum mechanics is actually needed at all to account for consciousness – unlike other biological phenomena that we have considered in this book such as enzyme action or photosynthesis. But is it likely that the strange features of quantum mechanics we have discovered to be involved in so many crucial phenomena of life are excluded from its most mysterious product, consciousness? We will leave the reader to decide. The scheme outlined above, involving quantum coherent ion channels and EM fields, is certainly speculative, but it does at least provide a plausible link between the quantum and classical realms in the brain.

So with this *in mind*, let us return once again to that dark cave in the South of France to complete the chain of events from brain to hand as our artist stands poised before the wall watching the torchlight flicker over its grey contours. Some play of the light and rock brings the image of a bison to her conscious mind. This is sufficient to create an idea in her head, perhaps instantiated as a fluctuation of her brain's EM field, that flips open up clusters of coherent ion channels in lots of separated neurons, causing them to fire synchronously. The synchronous nerve signals fire action potentials throughout her brain and, via synaptic connections, initiate a train of signals that travels down her spine and, via nerve–nerve junctions, to the motor nerves that discharge their packets of neurotransmitters into the neuromuscular junctions that are attached to the muscles of her arm. Those muscles contract to generate the coordinated motion of her hand that sweeps across the cave wall, depositing a line of charcoal on the rock in the shape of a bison. And, perhaps more importantly, she perceives that she initiated the action *because* of an idea in her conscious mind. She is not a zombie.

Thirty thousand years later, Jean-Marie Chauvet shines a torch on that same cave wall and the idea that came to life within the brain of that long-dead artist is once again flickering through the neurons of a conscious human mind.

9

How life began

... if (and oh! what a big if!) we could conceive [of] some warm
little pond, with all sorts of ammonia and phosphoric salts, light,
heat, electricity etc. present, that a protein compound was
chemically formed ready to undergo still more complex changes ...
Charles Darwin, letter to Joseph Hooker, 1871

G REENLAND IS not particularly green. Some time around AD 982,
a Danish Viking known as Erik the Red fled a charge of murder
by sailing westward from Iceland and discovered the island. He wasn't
the first: it had already been discovered several times by stone age
people who arrived from eastern Canada as early as 2500 BCE. But
Greenland's is a harsh and unforgiving environment, and those earlier cultures vanished, leaving only faint traces. Erik hoped to fare
better, arriving during the so-called Medieval Warm Period when
conditions were more clement; he therefore gave the island its current name, trusting that its promise of verdant pastures would lure
his fellow countrymen westward. The ploy evidently worked, because
a colony of several thousand was soon established and, initially at
least, appeared to thrive. But as the warm period waned Greenland
returned to climatic conditions more typical of the North Atlantic,
and the central icecap grew to cover 80 per cent of the island's
landmass. With the weather turning increasingly fierce, the island-
ers struggled to sustain their Scandinavian farming system in the

shallow soil of the thin coastal strip, and both crop yields and live-stock dwindled.

Ironically, at about the same time as the Viking colony was failing, another wave of immigrants, the Inuit (Eskimos), were making a living in the north of the island with a sophisticated fishing and hunting technology that was well adapted to the local conditions. The Vikings could have been saved if they had borrowed survival strategies from the Inuit, but the only record we have of contact between the two peoples is the remark from a Viking settler that the Inuit bleed a lot when stabbed; an observation that hardly indicates a willingness to learn from their northern neighbours. The result was that some time in the late fifteenth century the Viking colony collapsed, the last few inhabitants having apparently resorted to cannibalism.

However, the Danes never forgot about their western outpost and in the early eighteenth century an expedition was sent out to renew ties with the settlers. They found only abandoned homesteads and graveyards, but the visit did lead to the establishment of a more successful colony that, along with the native Inuit, eventually became the modern state of Greenland. The economy of Greenland today has grown from its Inuit roots, depending largely on fishing, but the potential mineral wealth of the island has been increasingly recognized. In the 1960s, the Danish Geological Survey of Greenland hired a young New Zealand-born geologist named Vic McGregor to conduct a geological study of the south-west corner of the island near its capital, Godthaab (now renamed Nuuk).

McGregor spent several years travelling through the fjord-riven region in a tiny, partly open boat, just big enough for himself, two local crew and the occasional guest, all crammed in among camping, hunting and fishing tools – not dissimilar from the kit of those early Inuit colonists – and geological equipment. Using standard techniques of stratigraphy, he concluded that the rocks in the area had been laid down in ten successive layers, of which the oldest and deepest was likely to be 'very old indeed' – perhaps even more than three billion years old.

In the early 1970s, McGregor sent a sample of his ancient rock to the Oxford laboratory of Stephen Moorbath, a scientist who had established a reputation for radiometric dating of rocks. The method depends on measuring the ratio of radioactive isotopes and their decay products. For example, uranium 238 decays with a half-life of 4.5 billion years (through a chain of nuclides, eventually into a stable isotope of lead); so, as the earth is about four billion years old, the concentration of natural uranium in a rock will take the entire age of the earth to drop by half. By measuring the ratio of these isotopes in any sample of rock, scientists can therefore calculate how long it is since those rocks were laid down; and it was these techniques which Stephen Moorbath used in 1970 to analyse a sample of a type of rock called *gneiss*, which McGregor had chipped out of the south-west Greenland coastal region known as Amîtsoq. Amazingly, he discovered that the gneiss contained proportionally more lead than any terrestrial ore or rock ever reported. The finding of very high levels of lead meant the Amîtsoq gneiss rock was, as McGregor had guessed, 'very old indeed'; at least 3.7 billion years old, older than any rock previously found on earth.

Moorbath was so struck by the discovery that he then joined McGregor on several expeditions to Greenland. In 1971, the two of them decided to visit the remote and virtually unexplored Isua region on the edge of the inland ice sheet (see figure 9.1). They first had to sail in McGregor's tiny boat up to the head of the iceberg-packed Godthaab Fjord, where the Viking settlers had eked out their precarious living in the middle ages. They were then picked up by a helicopter belonging to a local mining company that was also interested in the region, which aerial magnetic surveys had suggested was potentially rich in iron ore. The scientists discovered that within the local Isua greenstone there were many pillow-shaped masses of rock, known as basaltic pillow lavas, which had been formed by volcanic lava extruding directly into seawater: so-called *mud volcanoes*. These rocks were again dated back to at least 3.7 billion years ago. The finding clearly demonstrated that the earth had liquid warm

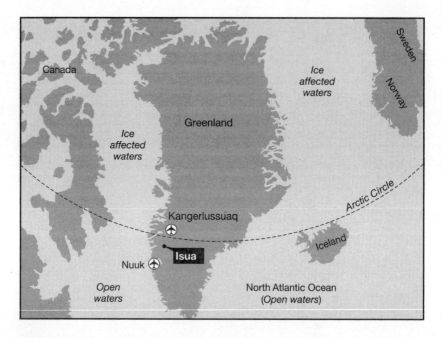

Figure 9.1: Map of Greenland, showing location of Isua.

oceans not long after its formation,* with mud volcanoes (figure 9.2) bubbling out of hydrothermal vents at the bottom of a shallow sea.

However, the real surprise came when Minik Rosing, a researcher from the Geological Museum in Copenhagen, measured the ratio of carbon isotopes in the Isua greenstone. The rocks contain about 0.4 per cent carbon, and when the respective ratios of the two isotopes ^{13}C and ^{12}C were measured, it was found that the amount of the heavier and rarer ^{13}C in the rocks was much lower than expected. Inorganic sources of carbon, such as atmospheric carbon dioxide, have about 1 per cent ^{13}C, but photosynthesis prefers to incorporate the lighter ^{12}C isotope into plant and microbial biomass, so a low level of ^{13}C is generally an indicator of the presence of organic material. These results suggested that within the warm waters surrounding the

---

* The earth is thought to have condensed out of solar remnants about 4.5 billion years ago, but only to have formed a solid crust about half a billion years later.

Figure 9.2: A modern mud volcano in Trinidad. Could the first life on earth have bubbled out of a similar mud volcano, leaving its traces in Isua greenstone? Photo: Michael C. Rygel via Wikipedia Commons.

Isua mud volcanoes 3.7 billion years ago there lived organisms that, like modern-day plants, were capturing carbon from carbon dioxide, either from the atmosphere or dissolved in water, and using it to construct all the carbon-based compounds that make up their cells.

The Isua rocks theory remains controversial, and many scientists are not convinced that the low levels of ^{13}C found there necessarily imply so early a presence of living organisms. Much of the scepticism derives from the fact that 3.8 billion years ago the earth was in the throes of what is known as the 'Late Heavy Bombardment', suffering regular impacts from asteroids and comets with energies sufficient to vaporize any surface water and presumably also to sterilize the oceans. Discovery of fossils of any such ancient photosynthesizing organisms would of course clinch the case, but the Isua rocks have been severely deformed over the millennia and any such fossils would be unrecognizable. We have to skip forward at least several hundred million years before proof of the existence of life is clearly present in the form of recognizable fossils of ancient microbes.

Notwithstanding the lack of conclusive evidence, many believe that the Isua isotope data provide the earliest indications of life on earth; and the Isua mud volcanoes would certainly have provided an ideal environment for the emergence of life, with their warm alkaline waters spouting from thermal vents. They would have been rich in dissolved inorganic carbonates, and the extruded snake-like *serpentine* rocks, which are highly porous, would have been riddled with billions of tiny cavities, each of which could have been a microenvironment capable of concentrating and stabilizing tiny amounts of organic compounds. Perhaps life really did first become green in the mud of Greenland. The question is: how?

The gunk problem

The three greatest mysteries in science are generally reckoned to be the origin of the universe, the origin of life and the origin of consciousness. Quantum mechanics is intimately involved in the first, and we have already discussed its possible connection to the third; as we will soon discover, it may also help to account for the second mystery. But we should first examine whether non-quantum explanations are able to provide a complete account of the origin of life.

The scientists, philosophers and theologians who have for centuries pondered the origin of life have come up with a rich variety of theories to explain it, ranging from divine creation to the seeding of our planet from space in the so-called *panspermia* theory. A more rigorously scientific approach was initiated in the nineteenth century by scientists such as Charles Darwin who proposed that chemical processes taking place in some 'warm little pond' may have led to the creation of living material. The formal scientific theory that built upon Darwin's speculations was put forward separately and independently by a Russian, Alexander Oparin, and an Englishman, J. B. S. Haldane, at the beginning of the twentieth century and is now generally known as the Oparin–Haldane hypothesis. Both proposed that the atmosphere of the early earth was rich in hydrogen, methane and water vapour which, when exposed to lightning, solar radiation or volcanic heat, combined to form a mixture of simple organic compounds. They proposed that these compounds then accumulated in the primordial ocean to form a warm, dilute organic soup, which swilled around in the water for millions of years, perhaps flowing over the Isua mud volcanoes, until some chance combination of its constituents eventually yielded a new molecule with an extraordinary property: the ability to replicate itself.

Haldane and Oparin proposed that the emergence of this *primordial replicator* was the key event that led to the origin of life as we know it. Its subsequent success would still have been subject to Darwinian natural selection. As a very simple entity, the replicator would have generated many errors or mutations in its replication. These mutant replicators would have then competed with non-mutated forms for the chemical materials from which to build more replicators. Those that were most successful would then have left the greatest number of descendants, and a molecular process of Darwinian natural selection would have taken hold to drive the swarm of replicators towards greater efficiency and greater complexity. Replicators that captured accessory molecules, such as peptides, that enzymatically catalysed their replication would have gained an advantage, and some may even have become enclosed within vesicles (tiny fluid- or air-filled sacs) bounded by fatty membranes, as today's living cells are, that protected them from the vagaries of their exterior environment. Once enclosed, the interior of the cell would then be able to support *bio*chemical transformations – its *metabolism* – to make its own biomolecules and prevent them from leaking out. With the ability to maintain and sustain its internal state while keeping it separated from its environment, the first living cell would have been born.

The Oparin–Haldane hypothesis provided a scientific framework within which to understand how life could have originated on earth. Yet for several decades the theory went untested – until two American chemists took an interest.

By the 1950s, Harold Urey was a distinguished but controversial scientist. He had been awarded the Nobel Prize in chemistry in 1934 for discovering deuterium, the isotope of hydrogen which, as you may remember from chapter 3, was used to study the kinetic isotope effect in enzymes and thereby demonstrate that their activity involves quantum tunnelling. Urey's expertise in the purification of isotopes led to his appointment in 1941 as head of the uranium enrichment part of the Manhattan Project, which was attempting to develop the atomic bomb. However, Urey became disillusioned with the Manhattan Project's aims and the secrecy in which it operated, and later attempted to dissuade the US President, Harry S. Truman, from dropping the bomb on Japan. After Hiroshima and Nagasaki, Urey wrote an article for the popular *Collier's* magazine, entitled 'I'm a frightened man', warning of the dangers posed by atomic weapons. From his post at the University of Chicago he also actively opposed McCarthy's anti-communist 'witch hunts' of the 1950s, writing letters to President Truman in support of Julius and Ethel Rosenberg, who were tried for espionage and eventually executed for passing atomic secrets to the Soviets.

Stanley Miller, the other American chemist involved in testing the Oparin–Haldane hypothesis, joined the University of Chicago as a PhD student in 1951, working initially on the problem of the nucleosynthesis of elements inside stars, under the guidance of the scientist known as the 'father of the hydrogen bomb', Edward Teller. Miller's life changed when in October 1951 he attended a lecture given by Harold Urey on the origin of life, in which Urey discussed the feasibility of the Oparin–Haldane scenario and suggested that someone should do the experiments. Fascinated, Miller transferred from Teller's to Urey's lab and set about persuading Urey to become his PhD mentor and to allow him to carry out the experiments. Urey was initially sceptical about his enthusiastic student's plans to put the Oparin–Haldane theory to the test: it might, he reckoned, take millions of years for inorganic chemical reactions to generate a sufficient number of organic molecules to be detected, while Miller had just three years to get his PhD! Nevertheless, Urey was prepared to give him the space and resources he needed for six months to a year. That way, if the experiments were not going anywhere, Miller would have time to switch to a safer research project.

In his attempt to replicate the conditions in which life originated on the early earth, Miller simulated the primordial atmosphere by simply filling a bottle with water, to simulate the ocean, topped up with the gases that he thought would have been present in the atmosphere: methane, hydrogen, ammonia and water vapour. He then simulated lightning by igniting the mixture with electric sparks. To Miller's surprise, and to the general astonishment of the scientific world, he discovered that after only a week of sparking his primordial atmosphere the bottle contained significant quantities of amino acids, the building blocks of proteins. The paper describing this experiment was published in the journal *Science* in 1953[1] – with Miller as sole author, Harold Urey having adopted the highly unusual position of insisting that his PhD student gain full credit for the discovery.

The Miller–Urey experiment – as it is generally known today despite Urey's unselfish gesture – was hailed as the first step in the creation of life in the laboratory, and remains a landmark in biology. Although no self-replicating molecules were generated, it was generally believed that Miller's 'primordial' soup of amino acids would have polymerized to form peptides and complex proteins and, given enough time and a sufficiently large ocean, eventually yield the Oparin–Haldane replicators.

Since the 1950s the Miller–Urey experiment has been repeated in many different ways by scores of scientists using different mixtures of chemicals, gases and energy sources to generate not only amino acids, but sugars and even small quantities of nucleic acids. And yet here we are, more than half a century later, with no laboratory-created primordial soup having yet yielded an Oparin–Haldane primordial replicator. To understand why, we need to look more closely at Miller's experiments.

The first issue is the complexity of the chemical mixture that Miller generated. Much of the organic material produced was in the form of a complex tar, of the kind familiar to organic chemists who often see such substances whenever their complex chemical synthesis procedures are not strictly controlled and so lots of wrong products are made. In fact, it is easy to produce a similar tar in the comfort of your own kitchen just by burning the dinner: that blackish-brown gunk that is so hard to remove from the bottom of your pan is rather similar in composition to Miller's tar. The problem with such chemical mixtures is that it is notoriously difficult to produce anything more than this tar-like gunk from them. In chemical terms, they are not what is called 'productive', because they are so complex that any specific chemical, such as an amino acid, tends to react with so many other different compounds that it then gets lost in a forest of inconsequential chemical reactions. Millions of cooks, and thousands of undergraduate chemistry students, have been producing such organic gunk for centuries, resulting in little more than a tough washing-up task.

From gunk to cells

Imagine trying to make a primordial soup by scraping all the gunk off the bottom of all the burnt pots in the entire world and then dissolving all those trillions of complex organic molecules into an ocean-sized volume of water. Now add a few Greenland mud volcanoes as your source of energy, and perhaps the spark of lightning, and stir. How long would you have to stir your soup before you created life? A million years? A hundred million years? A hundred *billion* years?

Even the simplest life is, much like this chemical gunk, extraordinarily complex. Unlike gunk, however, it is also highly organized.

The problem with using gunk as the starting material for generating organized life is that the random thermodynamic forces that were available in the primordial earth – the billiard-ball-like molecular motions that we discussed in chapter 2 – tend to destroy order rather than create it. You throw a chicken into the pot, heat it up and stir it, and make chicken soup. No one has ever poured a can of soup into a pot and made a chicken.

Of course, life didn't start with chickens (or eggs). The most basic self-replicating organisms alive today are bacteria, which are far simpler than any bird.* The simplest is called a mycoplasma (the bacterium that was the subject of Craig Venter's synthetic life experiment); but even these creatures are extremely complex life forms. Their genome encodes nearly five hundred genes, which produce a similar number of highly complex proteins that, as enzymes, make lipids, sugars, DNA, RNA, the cell membrane, its chromosome and a thousand other structures, each far more intricate than your car engine. And, in reality, mycoplasma is actually a bit of a bacterial wimp as it cannot survive on its own and must obtain many of its biomolecules from its host: it is a parasite and, as such, would be unable to survive in any realistic primordial soup. A more likely candidate would be another single-celled organism called a cyanobacterium that is able to photosynthesize to make all its own biochemicals. If present on the early earth, these cyano-bacteria would have been a potential source of those low levels of ^{13}C detected in the 3.7-billion-year-old Isua rocks in Greenland. But this bacterium is much more complex than a mycoplasma, with a genome encoding nearly *two thousand* genes. How long would you have to stir your ocean of primordial soup to make a cyanobacterium?

The British astronomer who coined the term 'Big Bang', Sir Fred Hoyle, had an interest in the origins of life that lasted throughout his own lifetime. The probability of random chemical processes coming

---

* This excludes viruses, which can only replicate with the help of a living cell.

together to generate life, he said, was as likely as a tornado blowing through a junkyard and assembling a jumbo jet by chance. The point he was making so vividly was that cellular life, as we know it today, is just too complex and organized to have arisen by chance alone; it must have been preceded by simpler self-replicators.

The RNA world

So what were those early self-replicators like? And how did they work? As none survive today, presumably because they have been out-competed into extinction by their more successful descendants, their nature is mostly educated guesswork. One approach is to extrapolate backwards from the simplest life forms alive today to imagine a much simpler self-replicator, a kind of stripped-down bacterium that may have been the precursor, billions of years ago, of all life on earth.

The problem is that it's not possible to dissect simpler self-replicators out of living cells because none of the components of cells are capable of self-replication by themselves. DNA genes don't replicate themselves; that is the job of the DNA polymerase enzymes. In turn, those enzymes don't replicate themselves, for they need to be first encoded within DNA and RNA strands.

RNA will play an important role in this chapter, so it may be useful to recall what it is and what it does. RNA is DNA's simpler chemical cousin, and it comes as a single-stranded helix compared with DNA's double helix. Despite this difference, RNA has more or less the same genetic information coding capacity as its more famous cousin – it just doesn't have the complementary copy of that information. And, just like DNA, its genetic information is written in four different genetic letters, so genes can be encoded in RNA just as they can be in DNA. Indeed, many viruses, such as the influenza virus, possess RNA genomes, rather than DNA genomes. But in living cells such as bacteria, animal or plant cells, RNA performs a role distinct from DNA: the genetic information written into DNA is first copied into RNA in the gene-reading process that we discussed in chapter 7. And since, unlike the relatively massive and immobile DNA chromosome, shorter RNA strings are free to move around the cell, they can carry the genetic message of genes from the chromosome to the protein synthesis machinery. Here the RNA sequence is read and translated into the sequences of amino acids that go into proteins, such as enzymes. So, in modern cells at least, RNA is a key intermediary between the genetic code written in DNA and the proteins that go on to make all the other components of our cells.

Returning then to our origin-of-life problem, then, although a living cell as a whole is a self-replicating entity, its individual components are not; just as a woman is a self-replicator (with a little 'help'), but her heart or liver is not. This creates a problem when trying to extrapolate backwards from today's complex cellular life to its much simpler non-cellular ancestor. If you put it another way, the question becomes: which came first: the DNA gene, the RNA, or the enzyme? If DNA or RNA came first, then what made them? If the enzyme came first, then how was it encoded?

One possible solution was provided by the American biochemist Thomas Cech, who discovered in 1982 that as well as encoding genetic information, some RNA molecules could take on the job of enzymes to catalyse reactions (work for which he shared the 1989 Nobel Prize in chemistry with Sidney Altman). The first examples of these *ribozymes*, as they are known, were found in the genes of tiny single-celled organisms called *Tetrahymena*, which is a type of protozoan found in freshwater ponds; but ribozymes have since been found to play a role in all living cells. Their discovery was quickly seized upon as a possible way out of the chicken-and-egg origin of life conundrum. The *RNA world hypothesis*, as it came to be known, proposes that primordial chemical synthesis resulted in the generation of an RNA molecule that could act as both gene and enzyme, and thus could both encode its own structure (like DNA) and make copies of itself (like enzymes) out of the biochemicals available in the primordial soup. This copying process would initially have been very hit-and-miss, giving rise to lots of mutant versions that would have competed against each other in the molecular Darwinian competition envisaged earlier. Over the course of time, those RNA replicators would have recruited proteins to improve their replication efficiency, leading to DNA and eventually the first living cell.

The idea that a world of self-replicating RNA molecules preceded the emergence of DNA and cells is now almost dogma in origin-of-life research. Ribozymes have been shown to be able to perform all the key reactions expected of any self-replicating molecule. For example, one class of ribozymes can join two RNA molecules together, whereas another can break them apart. Yet another form of ribozyme can make copies of short strings (just a handful of bases long) of RNA bases. From these simple activities we can imagine a more complex ribozyme able to catalyse the complete set of reactions necessary for self-replication. Once self-replication kicks in, then so too does natural selection; so the RNA world would have been set on a competitive path that led eventually, or so it is argued, to the first living cell.

There are, however, several problems with this scenario. Although simple biochemical reactions may be catalysed by ribozymes, self-replication of a ribozyme is a far more complex process involving recognition by the ribozyme of the sequence of its own bases, identification of identical chemicals in the ribozyme's environment, and assembly of those chemicals in the correct sequence to make a replica of itself. This is a tall order even for proteins having the luxury of living within cells packed full of the correct biochemicals, so it is even harder to see how ribozymes surviving in the messy and gunky primordial soup could achieve this feat. To date, no one has discovered or succeeded in making a ribozyme that can undertake such a complex task, even in the laboratory.

There is also the more fundamental problem of how to make the RNA molecules themselves in the primordial soup. The molecule is made of three pieces: the RNA base that encodes its genetic information (just as DNA bases encode the DNA's genetic information), a phosphate group and a sugar called ribose. Although some success has been achieved in devising plausible chemical reactions that might have made the RNA bases and phosphate components in the primordial soup, the most credible reaction that yields ribose also produces a plethora of other sugars. There is no known non-biological mechanism by which the ribose sugar can be generated on its own. And even if the ribose sugar were made, putting all three components together correctly is itself a formidable task. When plausible forms of the three components of RNA are brought together, they just combine in arbitrary ways to form the inevitable primordial gunk. Chemists avoid this problem by using special forms of bases whose chemical groups have been modified to avoid those unwanted side reactions – but this is cheating; and, in any case, the 'activated' bases are even more unlikely to have been formed in primordial conditions than the original RNA bases.

However, chemists are able to synthesize the RNA bases from simple chemicals by going through a very complex series of carefully controlled reactions in which each desired product from one reaction is isolated and purified before taking it on to the next reaction. The Scottish chemist Graham Cairns-Smith estimated that there are about 140 steps necessary for the synthesis of an RNA base from simple organic compounds likely to have been present in the primordial soup.[2] For each step there is a minimum of about six alternative reactions that need to be avoided. This makes the chemical synthesis easy to visualize, for you can conceive of each molecule as a kind of molecular die, with each step corresponding to a throw where the number six represents generating the correct product and any other number indicates that the wrong product has been made. So, the odds of any starting molecule eventually being converted into RNA is equivalent to throwing a six 140 times in a row.

Of course, chemists improve these stupendous odds by carefully controlling each step, but the prebiotic world would have had to rely on chance alone. Perhaps the sun came out at just the right time to evaporate a little pool of chemicals surrounding a mud volcano? Or perhaps the mud volcano erupted to add water and a little sulphur to create another set of compounds? Perhaps a lightning storm stirred up the mix and accelerated a few more chemical changes with the input of electrical energy? The questions could go on and on; but it's easy enough to estimate the probability that, relying on chance alone, each of the 140 necessary steps would have yielded the right one of six possible products: it is one in 6^{140} (roughly, 10^{109}). To have a statistical chance of making RNA by purely random processes you would need at least this number of starting molecules in your primordial soup. But 10^{109} is a far bigger number than even the number of fundamental particles in the entire visible universe (about 10^{80}). The earth simply did not have enough molecules, or sufficient time, to make significant quantities of RNA in those millions of years between its formation and the emergence of life at the time suggested by the Isua rocks.

Nevertheless, imagine that the synthesis of significant quantities of RNA *did* happen, through some as yet undiscovered chemical process. We now have to overcome the equally daunting problem of stringing the four different RNA bases (equivalent, you'll remember, to those four letters of the DNA code, A, G, C and T) together in just the right sequence to make a ribozyme capable of self-replication. Most ribozymes are RNA strings at least a hundred bases long. At each position in the string one of the four bases must be present, so there are 4^{100} (or 10^{60}), different ways to put together a string of RNA 100 bases long. How likely is it that the random jumbling together of RNA bases will generate just the right sequence along the length of the string to make a self-replicating ribozyme?

Since we seem to be having such fun with big numbers, we can work it out. It turns out that 4^{100} individual strings of RNA 100 bases long would have a combined mass of 10^{50} kilograms. So this is how much we would need, in order to have a single copy of most strings and therefore a reasonable chance that one of them would have all its bases arranged correctly to be a self-replicator. However, the entire mass of the Milky Way galaxy is estimated to be approximately 10^{42} kilograms.

Clearly, we cannot rely on pure chance alone.

Of course, there may not be just one arrangement among the 4^{100} possible 100 base-long RNA strings that would act as a self-replicator. There may be many more. There could even be trillions of possible replicators that can be formed out of RNA strings 100 bases long. Perhaps self-replicating RNA is actually quite common, and we only need a million molecules to have some chance of forming a self-replicator. The problem with this argument is that it is just that: an argument. Despite many attempts, no one has ever made a single self-replicating RNA (or DNA, or protein), or observed one in nature. This is not so surprising when you consider what a challenging job self-replication is. In today's world it takes an entire living cell to achieve this feat. Could it have been done with a far simpler system billions of years ago? Surely it must have, or we wouldn't be here contemplating the problem today. But how this was achieved before cells evolved is far from clear.

Given the difficulties of identifying biological self-replicators, we might gain insight by asking a more general question: how easy is self-replication in any system? Modern technology has provided us with lots of machines that can replicate stuff, from photocopying machines to electronic computers to 3D printers. Can any of these devices make a copy of itself? Probably the closest is a 3D printer such as one of the RepRap (short for *Replicating Rapid prototyper*) printers that are the brainchild of Adrian Bowyer at the University of Bath in the UK. These machines can print their own components, which can then be assembled to make another RepRap 3D printer.

Well, not quite. The machine only prints in plastic, but its own frame is made of metal, as are most of its electrical components. So it is only the plastic parts that it can replicate; and these have to be manually assembled with additional parts to make a new printer. The vision of the designers is to make self-replicating RepRap printers (there are several alternative designs) freely available for the benefit of everyone. But at the time of writing we are a long way from building a truly self-replicating machine.

So, if looking for self-replicating machines doesn't really help us in our quest to discover how easy or difficult self-replication is, can we eschew the material world entirely and explore this question within a computer, where those messy and hard-to-make chemicals can be replaced by the simple building blocks of the digital world: namely, the bits that can only have a value of either 1 or 0? A 'byte' of data, consisting of eight bits, represents a single character of text in a computer code and can be roughly equated with the unit of genetic code: a DNA or RNA base. We can now ask the question: among all the possible strings of bytes, how common are those that can replicate themselves in a computer?

Here we have a huge advantage, because self-replicating strings of bytes are actually quite common: we know them as computer viruses. These are relatively short computer programs that can infect our computers by persuading their CPU to make loads of copies. These computer viruses then hop into our emails to infect the computers of our friends and colleagues. So if we consider the computer memory as a kind of digital primordial soup, then computer viruses can be considered to be the digital equivalent of primordial self-replicators.

One of the simplest computer viruses, Tinba, is only 20 kilobytes long: very short compared to most computer programs. Yet Tinba successfully attacked the computers of large banks in 2012, burrowing into their browsers and stealing login data; so it was clearly a formidable self-replicator. While 20 kilobytes may be very short for a computer program, it nonetheless comprises a relatively long string of digital information as, with 8 bits in a byte, it corresponds to 160,000 bits of information. Since each bit can be in one of two states (0 or 1) we can easily calculate the probability of randomly generating particular strings of binary digits. For example, the chances of generating a particular three-bit string, say, 111, is $\frac{1}{2} \times \frac{1}{2} \times \frac{1}{2}$, or

1 chance in 2^3. Following the same mathematical logic, it follows that arriving by accident at a specific string 160,000 bits long, the length of Tinba, is 1 chance in $2^{120,000}$. This is a mind-bogglingly small number, and tells us that Tinba could not have arisen by chance alone.

Perhaps there are, just as we conjectured for RNA molecules, very many self-replicating codes out there that are far simpler than Tinba and that might have arisen by chance. But if that were the case, then surely a computer virus would, by now, have arisen spontaneously from all the zillions of gigabytes of computer code that are flowing through the internet every second. Most of these codes are after all just sequences of ones and zeros (think of all the images and movies that are being downloaded every second). These codes are all potentially functional in terms of instructing our CPUs to perform basic operations, such as to copy or to delete; yet all of the computer viruses that have ever infected anyone's computer show the unmistakable signature of human design. As far as we know, the vast stream of digital information that flows around the world every day has never spontaneously generated a computer virus. Even within the replication-friendly environment of a computer, self-replication is hard, and, so far as we know, it has never happened spontaneously.

So, can quantum mechanics help?

This excursion into the digital world exposes the essential problem in the quest for life's origin, which boils down to the nature of the search engine used to bring its necessary ingredients together in the correct configuration to form a self-replicator. Whatever chemicals were available in the primordial soup, they would have had to explore a huge space of possibilities to hit upon an exceedingly rare self-replicator. Could our problem be that we are confining the search routine to the rules of the classical world? You may remember from chapter 4 that the quantum theorists at MIT were initially highly sceptical of the *New York Times* report that plants and microbes were implementing a quantum search routine. But they eventually came round to the idea that photosynthetic systems were indeed implementing a quantum search strategy, called a quantum walk. Several researchers, ourselves included,[3] have explored the idea that the origin of life could similarly have involved some kind of quantum search scenario.

Imagine a tiny primordial pool enclosed within a pore of those serpentine rocks extruded from a mud volcano under the ancient Isua sea three and a half billion years ago, when Greenland's gneiss strata were being formed. Here is Darwin's 'warm little pond, with all sorts of ammonia and phosphoric salts, light, heat, electricity etc. present,' in which 'a protein compound . . . ready to undergo still more complex changes' might have formed. Now, further imagine that one 'protein compound' (it could just as easily be an RNA molecule), made by the kind of chemical processes that Stanley Miller discovered, is a kind of proto-enzyme (or ribozyme) that has some enzymatic activity, but is not yet a self-replicating molecule. Further imagine that some of the particles in this enzyme could move to different positions but are prevented from doing so by classical energy barriers. However, as we discussed in chapter 3, both electrons and protons are able to quantum tunnel through energy barriers that forbid their classical transfer, a feature that is crucial in enzyme action. In effect, the electron or proton exists on both sides of the barrier simultaneously. If we imagine this happening within our proto-enzymes, then we would expect the different configurations – finding the particle on either side of the energy barrier – to be associated with different enzyme activities, that is, abilities to accelerate different types of chemical reactions, perhaps including a self-replication reaction.

Just to make the numbers easy to work with, let us imagine there are a total of 64 protons and electrons within our imaginary proto-enzyme that are each capable of quantum tunnelling into any one of two different positions. The total amount of structural variation available to our imaginary proto-enzyme is still enormous: 2^{64} – an awful lot of possible configurations. Now imagine that just one of these configurations has what it takes to become a self-replicating enzyme. The question is: how easy is it to find the particular configuration that could lead to the emergence of life? Will the self-replicator ever be realized in our tiny warm pond?

Consider first the proto-enzyme as an entirely classical molecule unable to do any quantum tricks, such as superposition or tunnelling. The molecule must, at any given moment, be in just one of the possible 2^{64} different configurations, and the probability that this proto-enzyme will be a self-replicator is 1 divided by 2^{64} – an exceedingly small chance indeed. With overwhelming odds, the classical proto-enzyme will be stuck in one of the boring configurations that can't self-replicate.

Of course, molecules do change, as a result of general thermodynamic wear and tear, but in the classical world such change is relatively slow. For one molecule to change, the original arrangement of atoms must be dismantled and its constituent particles rearranged to form a new molecular configuration. As we discovered in chapter 3 with the long-lived dinosaur collagen, chemical changes can sometimes take place over geological timescales. Considered classically, our proto-enzyme would take a very long time to explore even a tiny fraction of those 2^{64} chemical configurations.

However, the situation is radically different if we consider the 64 key particles in the proto-enzyme to be electrons and protons that can tunnel between their alternative positions. Being a quantum system, the proto-enzyme can exist in all its possible configurations simultaneously as a quantum superposition. The reason for our choice of the number 64 above now becomes clearer; it is the same number we explored when we were using the Chinese emperor's chessboard blunder to illustrate the power of quantum computing in chapter 8, with the tunnelling particles taking the role of the squares on the board or qubits. Our proto-self-replicator could, if it survived long enough, act as a 64-qubit quantum computer; and we have already discovered how powerful such a device would be. Perhaps it can use its huge quantum computational resources to compute the answer to the question: what is the correct molecular configuration for a self-replicator? In this guise, the problem and its potential solution become clearer. Consider the proto-enzyme to be in such a quantum superposition, and the search problem of finding the one in 2^{64} possible structures that is the self-replicator becomes solvable.

There is a hitch, though. You will remember that qubits have to remain coherent and entangled in order to perform quantum computing. Once decoherence kicks in, the superposition of 2^{64} different states collapses and just one remains. Does this help? On the face of it, no, because the chance of the quantum superposition collapsing into the single self-replicating state is the same as before: a minuscule 1 divided by 2^{64}, the same as the chances of throwing heads 64 times in a row. But what happens next is where the quantum description diverges from its classical counterpart.

If a molecule is not behaving quantum mechanically and finds itself, as it almost certainly will, with the wrong arrangement of atoms that is unable to self-replicate, trying out a different configuration would have to involve the geologically slow process of dismantling and rearranging molecular bonds. But, after decoherence of the equivalent quantum molecule, each of the 64 electrons and protons of our proto-enzyme will, almost instantaneously, be ready to tunnel again into a superposition of both of their possible positions to re-establish the original quantum superposition of 2^{64} different configurations. In its 64-qubit state, the quantum proto-replicator molecule could repeat its search for self-replication in the quantum world continuously.

Decoherence will rapidly collapse the superposition once again; but this time the molecule will find itself in another of its 2^{64} different classical configurations. Once again, decoherence will collapse the superposition, and once again the system will find itself in another configuration; and this process will continue indefinitely. Essentially, in this relatively protected environment, the making and breaking of the quantum superposition state is a reversible process: the quantum coin is being continually tossed by the processes of superposition and decoherence, processes that are far more rapid than the classical making and breaking of chemical bonds.

But there is one event that will terminate the quantum coin-tossing. If the quantum proto-replicator molecule eventually collapses into a self-replicator state, it will start to replicate and, just as in the starving *E. coli* cells we discussed in chapter 7, replication will force the system to make an irreversible transition into the classical world. The quantum coin will have been irreversibly thrown, and the first self-replicator will have been born into the classical world. Of course, this replication will have to involve some sort of biochemical process within the molecule, or between it and its surroundings, that is distinctly different from those that took place before the proto-replicator arrangement was found. In other words, there needs to be a mechanism that anchors this special configuration in the classical world before it is lost and the molecule moves on to the next quantum arrangement.

What did the first self-replicator look like?

The proposition we have outlined above is, of course, speculative. But if the search for the first self-replicator was performed in the quantum rather than the classical world, it does at least potentially solve the self-replicator search problem.

In order for this scenario to work, the primordial biomolecule – the proto-self-replicator – must have been capable of exploring lots of different structures by the quantum tunnelling of its particles into different positions. Do we know what kind of molecules would be capable of such a trick? Well, to a certain extent we do. As we have already discovered, the electrons and protons in enzymes are held relatively loosely, which enables them to tunnel into different positions with ease. The protons in DNA and RNA are also capable of tunnelling, at least across the hydrogen bond. So we might imagine our primordial self-replicator to be something like a protein or RNA molecule that was loosely held together by hydrogen bonds and weak electronic bonds that allowed its particles – both protons and electrons – to travel freely through its structure to form a superposition of its trillions of different configurations.

Is there any evidence for such a scenario? Apoorva D. Patel, a physicist at the Centre for High Energy Physics at the Indian Institute of Science in Bangalore, is one of the world's experts on quantum algorithms – the software of quantum computers. Apoorva suggests that aspects of the genetic code (the sequences of DNA bases that code for one amino acid or another) betray its origin as a quantum code.[4] This is not the place to go into any technical detail (for this would take us too deeply into the mathematics of quantum information theory), but his idea should not come as such a surprise. In chapter 4 we saw how, in photosynthesis, the photon's energy is transferred to the reaction centre by following multiple pathways at once – a quantum random walk. Then, in chapter 8, we discussed the idea of quantum computation and whether life might make use of quantum algorithms to enhance the efficiency of certain biological processes. Similarly, origin-of-life scenarios that involve quantum mechanics, while speculative, are nothing more than an extension of these ideas: the possibility that quantum coherence in biology played the kind of role in the origin of life as it currently does in living cells.

Of course, any scenario involving quantum mechanics in the origin of life three billion years ago remains highly speculative. But, as we have discussed, even classical explanations of life's origin are beset with problems: it isn't easy to make life from scratch! By providing more efficient search strategies, quantum mechanics may have made the task of building a self-replicator a little easier. It almost certainly was not the whole story; but quantum mechanics could have made the emergence of life in those ancient Greenland rocks a lot more likely.

10
Quantum biology: life on the edge of a storm

'WEIRD' IS the adjective most frequently used to describe the field of quantum mechanics. And it is weird. Any theory that allows objects to pass through impenetrable barriers, to be in two places at once, or to possess 'spooky connections' cannot be described as ordinary. But in fact its mathematical framework is absolutely logical and consistent, and accurately describes the way the world is at the level of fundamental particles and forces. Quantum mechanics is thus the bedrock of physical reality. Discrete energy levels, wave–particle duality, coherence, entanglement and tunnelling aren't just interesting ideas relevant only to scientists working within rarefied physics laboratories. They are as real and as normal as Grandma's apple pie, and indeed are going on inside Grandma's apple pie. Quantum mechanics is normal. It is the world it describes that is weird.

But, as we have discovered, most of the counterintuitive features of matter at the quantum scale are washed away in the turbulent thermodynamic interiors of big objects by the process we call decoherence, leaving just our familiar classical world. So we can view physical reality as consisting of three levels (figure 10.1). On the surface are the macroscopic, everyday objects such as footballs, trains and planets, whose overall behaviour adheres to Newton's mechanical laws of motion involving such familiar concepts as speed, acceleration, momentum and forces. The middle layer is the thermodynamic layer that describes the behaviour of liquids and gases. Here, the same classical Newtonian rules apply; but, as Schrödinger pointed out and as we described in chapter 2, these underlying thermodynamic laws, which describe for example how a gas expands when heated or how a steam engine drives trains up hillsides, are based on the 'order from disorder' averaging of the disorderly billiard-ball-like jostling of trillions of atoms and molecules. The third and deepest level is the bedrock of reality: the quantum world. Here is where the behaviour of the atoms and molecules and the particles from which they are made obeys the precise and orderly rules of quantum, not classical, mechanics. However, most of the weird quantum stuff is generally invisible to us. It is only when we carefully observe individual molecules, as for example in the double-slit experiment, that we see the deeper, quantum laws. The behaviour they describe appears unfamiliar to us because we normally see reality through a decoherence filter that strips out all the weirdness from bigger objects.

Most living organisms are relatively large objects. Like trains, footballs and cannonballs, their overall motion adheres pretty well to Newtonian laws: a man fired out of a cannon has a similar trajectory as that of a cannonball. At a deeper level, the physiology of tissues and cells is also well described by the thermodynamic laws: the expansion and contraction of a lung is not so different from the expansion and contraction of a balloon. So at first glance you would tend to assume, and most scientists have assumed, that the quantum behaviour similarly gets washed away in robins, fish, dinosaurs, apple trees, butterflies and us, just as it does in other classical objects. But we have seen that this is not always true for life; its roots reach down from the Newtonian surface through the turbulent thermodynamic waters to penetrate the quantum bedrock, allowing life to harness coherence, superposition, tunnelling or entanglement (figure 10.1). The question we want to address in this final chapter is: how?

We have already explored part of the answer. Erwin Schrödinger pointed out more than sixty years ago that life is different from the inorganic world because it is structured and orderly even at a

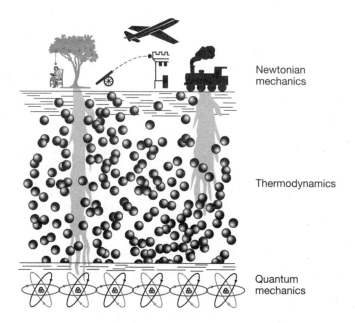

Newtonian
mechanics

Thermodynamics

Quantum
mechanics

Figure 10.1: The three strata of reality. The top layer is the visible world, filled with objects such as falling apples, cannonballs, steam trains and aeroplanes, whose motions are described by Newtonian mechanics. Lying beneath is the thermodynamic layer of billiard-ball-like particles whose motion is almost entirely random. This layer is responsible for generating the 'order from disorder' laws that govern the behaviour of objects such as steam engines. The next layer down is the layer of fundamental particles ruled by orderly quantum laws. The visible features of most of the objects that we see around us appear to be rooted in either the Newtonian or thermodynamic layers but living organisms have roots that penetrate right down to the quantum bedrock of reality.

molecular level. This order all the way down endows life with a kind of rigid leverage that connects the molecular to the macroscopic, such that quantum events taking place within individual biomolecules can have consequences for an entire organism: the kind of amplification from the quantum to the macroscopic asserted by that other quantum pioneer, Pascual Jordan.

Of course, when Schrödinger and Jordan were writing about biology nobody knew what a gene was made of, or how enzymes or photosynthesis worked. But half a century of intensive molecular biology research has provided us with an extraordinarily detailed map of the structure of biomolecules at the level of individual atoms in DNA or proteins. And, as we have discovered, the quantum pioneers' prescient insights have, rather belatedly, been vindicated. Photosystems, enzymes, respiratory chains and genes are structured right down to the position of individual particles, and their quantum motions do indeed make a difference to the respiration that keeps us alive, the enzymes that build our bodies or the photosynthesis that makes nearly all the biomass on our planet.

And yet many questions remain, principally concerning how life manages to maintain quantum coherence in the warm, wet sea of biomolecules within a living cell. Proteins or DNA are not steel-built machines with rigid parts, like the instruments used to detect quantum effects in physics laboratories; they are squishy, flexible structures that are constantly subjected to their own thermal vibrations as well as being continuously battered by the bumping of surrounding molecular billiard balls, a constant barrage of *molecular noise.** These random vibrations and collisions would be expected to shatter the delicate arrangement of atoms and molecules those particles need to maintain their quantum behaviour. How this coherence is preserved in biology remains a puzzle; but, as we will discover, it is one that is beginning to be unravelled to reveal fascinating insights into how life works; insights that might even be exploited to drive the quantum technologies of the future.

Good, good, good, good vibrations (bop bop)

Few popular science books require revision during their writing; but in this final chapter we will describe results that are emerging right now. Indeed, the science of quantum biology is moving so fast, on so many different fronts, that this book will inevitably be a little out of date by the time it is published. The biggest surprises to emerge from

---

* The term often used to describe incoherent molecular vibrations.

recent studies are new insights into how life copes with molecular vibrations or noise.

Some of the most exciting new results in this area are emerging from further studies of photosynthesis. You will remember from chapter 4 that microbes and plant leaves are packed full of chloroplasts filled with forests of chlorophyll pigment molecules, and that the first step in photosynthesis involves the capture of a photon of light by a pigment molecule and its conversion to an oscillating exciton that gets whisked through the chlorophyll forest to the reaction centre. You will also remember that the signature of coherence, quantum beating, was detected in this energy transport process – evidence that its near 100 per cent efficiency is thanks to excitons quantum walking their way to the reaction centre. But how excitons maintain their coherent wave-like behaviour while strolling through the molecularly noisy environment of a living cell has, until recently, been a puzzle. We have now discovered that the answer seems to be that living systems don't try to avoid molecular vibration; instead, they dance to its beat.

In chapter 4 we envisaged quantum coherence in photosynthesis as a kind of molecular version of an orchestra being 'in tune' and 'in time', with all the coherent pigment molecules playing to the same beat. But the problem the system has to overcome is that the inside of the cell is very noisy. This molecular orchestra is playing not in a quiet concert hall, but in something more like a busy city centre, amid a cacophony of molecular noise that disturbs each of the musicians so that their exciton oscillations are likely to be knocked out of tune, causing their delicate quantum coherence to be lost.

This challenge is familiar to physicists and engineers attempting to build devices such as quantum computers. They tend to use two main strategies to keep the noise at bay. First, whenever they can, they cool their systems down to very close to absolute zero. At these very low temperatures, the molecular vibrations are damped, which in turn subdues the molecular noise. Second, they shield their equipment within the molecular equivalent of a sound studio, thereby keeping any environmental noise at bay. There are no sound studios inside living cells, and plants and microbes live in hot environments, so how do photosystems maintain their tuneful quantum coherence for so long?

The answer appears to be that photosynthetic reaction centres exploit two varieties of molecular noise to maintain rather than destroy coherence. The first is a relatively weak and low-level noise, sometimes called *white noise*, which is rather like TV or radio static that is spread across all frequencies.* This white noise comes from the thermal molecular jostling of all the surrounding molecules, such as water or metal ions, that are packed inside living cells. The second kind, sometimes called *coloured noise*, is 'louder' and limited to certain frequencies, just as coloured (visible) light is limited to a narrow range of frequencies on the electromagnetic spectrum. The source of coloured noise is the vibrations of the larger molecular structures within the chloroplasts, such as the pigment (chlorophyll) molecules and the protein scaffolds that hold them in place, which are composed of strings of amino-acid beads that are bent and twisted into shapes suitable for housing pigment molecules. Their bends and twists are flexible and they can vibrate, but they do so only at certain frequencies, rather like the strings of a guitar. The pigment molecules themselves also have their own vibrational frequencies. These vibrations generate the coloured noise that, like a musical chord, is composed of just a few notes. Both white and coloured noise appear to be exploited by photosynthetic reaction systems to help shepherd the coherent exciton to the reaction centre.

A clue to how life exploits this type of molecular vibration was discovered independently by two groups in 2008–9. One was the then UK-based husband-and-wife team of Martin Plenio and Susana Huelga, who had long been interested in the effects of external 'noise' on the dynamics of quantum systems and so were not surprised when they heard about Graham Fleming's 2007 experiment on

---

* The amplitude of the vibrations is quite small, so they do not deliver much energy.

photosynthesis that we discussed in chapter 4. They quickly published several, now widely cited, papers providing a model of what they thought was going on.[1] They proposed that the noisy interior of a living cell might act to drive quantum dynamics and maintain quantum coherence in photosynthetic complexes and other biological systems rather than destroy it.

The other group, across the Atlantic, was that MIT-based quantum information team led by Seth Lloyd who initially thought quantum mechanics in plants was a crackpot idea. Together with colleagues from nearby Harvard University, Lloyd took a closer look at the algal photosynthetic complex in which Fleming and Engel had detected quantum beating.[2] They showed that transporting the quantum coherent exciton can be either retarded or assisted by environmental noise, depending on just how *loud* that noise is. If the system is too cold and quiet, then the exciton tends to oscillate aimlessly without actually getting anywhere in particular; whereas in a very hot and noisy environment something called the *quantum Zeno effect* kicks in, which retards quantum transport. Between these two extremes is a *Goldilocks zone* where vibrations are just right for quantum transport.

The quantum Zeno effect is named after the ancient Greek philosopher Zeno of Elea, who posed philosophical problems in the form of a set of paradoxes, one of which is known as the arrow paradox. Zeno considered an arrow in flight which, he argued, must inhabit a particular position in space for every instant of time. If the arrow could be glimpsed at that instant then it would be indistinguishable from a truly motionless arrow suspended in the same position. The paradox is that the flight of an arrow consists of a sequence of these frozen slices in time, with a motionless arrow at each point along its path. Yet when you put all the slices together, the arrow moves. So how can a sequence of such zero motions ever add up to real motion? The answer, we now know, is that a finite duration of time is not made up of a sequence of indivisible units of zero time. But that resolution had to wait until the invention of calculus in the seventeenth century, more than two thousand years after Zeno posed his puzzle. Nevertheless, Zeno's paradox survives, at least in name, in one of the most peculiar features of quantum mechanics. Quantum arrows really can be frozen in time by the act of observation.

In 1977, physicists at the University of Texas published a paper that showed how something akin to Zeno's arrow paradox can occur in the quantum world.[3] The quantum Zeno effect, as it came to be known, describes how continuous observations can prevent quantum events from happening. For example, a radioactive atom, if observed closely and continuously, will never decay – an effect often described in terms of the old adage 'the watched pot never boils'. Real pots do, of course, eventually boil; time only seems to slow down when you badly need a cup of tea. However, as Heisenberg pointed out, in the quantum realm the act of watching (measuring) inevitably alters the state of the thing that is being watched.

To see how Zeno's paradox is relevant to life, we will return to the energy transport step of photosynthesis. Let's imagine that a leaf has just picked up a solar photon and converted its energy to an exciton. Considered classically, the exciton is a particle that is localized in space and time. But, as the double-slit experiment revealed, quantum particles also possess a diffuse wave character that enables them to exist in multiple places simultaneously as a quantum superposition. It is the exciton's waviness that is essential for efficient quantum transport, for this enables it, like a water wave, to explore multiple paths simultaneously. But if its quantum waviness breaks on the molecularly noisy rocks of decoherence inside the leaf, then its waviness will be lost and it will become a localized particle stuck in a single position. The noise essentially acts as a kind of continuous measurement, and if it is very intense then decoherence will take place very quickly, before quantum coherence has a chance to help the exciton wave reach its destination. This is the quantum Zeno effect: constantly collapsing the quantum wave into the classical world.

When the MIT team estimated the influence of molecular noise/ vibrations in the bacterial photosynthetic complex, they discovered that quantum transport was optimal at temperatures around those at which microbes and plants perform photosynthesis. This perfect match between optimal transport efficiency and the kind of temperatures in which living organisms live is remarkable and, the team claim, suggests that three billion years of natural selection have fine-tuned the quantum-level evolutionary engineering of exciton transport to optimize the most important biochemical reaction in the biosphere. As they argue in a later paper, 'natural selection tends to drive quantum systems to the degree of quantum coherence that is "just right" for attaining maximum efficiency'.[4]

However, good molecular vibrations are not just limited to the white noise variety. 'Coloured' noise, generated by a limited set of vibrations of the chlorophyll molecules themselves, or even the surrounding proteins, is now also thought to play a key role in keeping decoherence at bay. If we imagine the white thermal noise as a molecular version of the static on a badly tuned radio, then the good vibrations of coloured noise are akin to a simple beat like the Beach Boys' 'bop bop' in their song 'Good Vibrations'. But remember that the exciton also behaves in a wave-like manner to generate those coherent quantum beats that Graham Fleming's group detected. Two recent papers from Martin Plenio's group at the University of Ulm in Germany in 2012 and 2013 demonstrated that if the oscillation of the exciton and the oscillations of the surrounding proteins – the coloured noise – are beating to the same drum then, when the coherent exciton gets knocked out of tune by the white noise, it can be knocked back into tune by the protein oscillations.[5] Indeed, in a 2014 *Nature* paper, Alexandra Olaya-Castro at University College London showed in a beautiful theoretical study that the exciton and the molecular vibrations – the coloured noise – share a single quantum of energy in a way that simply cannot be explained without recourse to quantum mechanics.[6]

To fully appreciate the contributions of the two kinds of molecular noise to exciton transport, let us return to a musical metaphor once again and imagine the photosystem is an orchestra, with the various instruments playing the role of the pigment molecules, and the exciton is a musical tune. We imagine that the music opens with a violin solo, representing the pigment molecule that captures the photon and converts its energy into a vibrating exciton. The music of the exciton is then picked up by the other string instruments, then the wind instruments, and eventually reaches the percussion, whose rhythm plays the role of the reaction centre. We will further imagine that this music is playing in a theatre packed with an audience who will provide the white noise of crisp packets being opened, chairs being shuffled, coughs and sneezes. The conductor will be the coloured noise.

Let's first imagine we have arrived on a very rowdy night, with the audience making such a racket that the musicians cannot hear either themselves or their colleagues. In all the hubbub, the first violin begins the piece but the other musicians cannot hear it and so are unable to pick up the melody. This is the quantum Zeno scenario, where too much noise is preventing quantum transport. However, at very low levels of noise, say in an empty theatre without any audience present, the musicians are only listening to each other, so they all pick up the first melody, like a tune that you can't get out of your head, and keep playing it. This is the opposite scenario of too much quantum coherence, where the exciton remains oscillating throughout the whole system but doesn't end up anywhere in particular.

In the Goldilocks zone, with just the right amount of noise delivered by a self-controlled audience, the disturbance is sufficient to jog the musicians out of their monotonous repetition to play the full score with all its dynamics. Some of the instruments are still knocked into a different beat when an occasional crisp bag is popped by a rowdy spectator, but, with a wave of his baton, the conductor is able to bring them back into sync to deliver the music of photosynthesis.

Reflections on the motive force of life

In chapter 2 we peered inside a steam engine to discover that its motive force involved capturing the random motion of the sea of billiard-ball-like molecules and directing the molecular turbulence towards driving the piston within the cylinder. We then asked whether life can be entirely accounted for by the same 'order from disorder' thermodynamic principle that drives steam engines. Is life just an elaborate steam engine?

Many scientists are convinced that it is, but in a subtle way that needs a little elaboration. Complexity theory studies the tendency of certain forms of random chaotic motion to generate order through the phenomenon of *self-organization*. For example, as we have already discussed, the molecules within liquids are moving entirely chaotically, yet when your bathtub is draining the water spontaneously flows around the plughole in an orderly clockwise or anticlockwise direction. This macroscopic order can also be seen in the patterns of convection flow in a heated pot of water, in hurricanes, tornadoes, the red spot on Jupiter and many other natural phenomena. Self-organization is also involved in several biological phenomena, such as the swarming behaviour of birds, fish or insects, or in the pattern of stripes of a zebra, or in the complex fractal structure of some leaves.

What is remarkable about all these systems is that the macroscopic order we can see is not reflected at the molecular level. If you had a very powerful microscope that could reveal the individual molecules that were flowing down your plughole you might be surprised to see that their motions are nearly entirely random, with just a very slight bias from randomness in a clockwise or anticlockwise direction. At a molecular level, there is only chaos – but chaos with a slight bias, that can generate order at a macroscopic level: order from chaos, as this principle is sometimes termed.[7]

Order from chaos is conceptually quite similar to Erwin Schröding-er's 'order from disorder', which, as we have already described, lies behind the motive force of steam engines. But, as we have discovered, life is different. Although there is plenty of disorderly molecular motion inside living cells, the real action of life is a tightly choreographed motion of fundamental particles within enzymes, photosynthetic systems, DNA and elsewhere. Life has built-in order at a microscopic level; and so 'order from chaos' cannot be the only explanation for life's fundamental distinguishing features. Life is *nothing* like a steam train.

However, recent research suggests that life may operate along the lines of a quantum version of the steam engine.

The principle of how steam engines work was first outlined in the nineteenth century by a Frenchman, Sadi Carnot. He was the son of Napoleon's minister of war, Lazare Carnot, who obtained a commission in the engineer corps of Louis XVI's army. After the king was deposed, Lazare Carnot did not, like many of his aristocratic colleagues, flee the country, but instead joined the revolution; and, as war minister, he was largely responsible for creating the French revolutionary army that repelled the Prussian invasion. But as well as being a brilliant military strategist, Lazare was also a mathematician, a lover of music and poetry (he named his son after the medieval Persian poet Saadi Shirazi), and an engineer; he wrote a book on how machines convert one form of energy into another.

Sadi exhibited some of his father's revolutionary and nationalistic fervour, taking part in the defence of Paris as a student in 1814 when the city was once again besieged by the Prussians. He also demonstrated some of his father's engineering insight, writing a remarkable book entitled *Reflections on the Motive Force of Fire* (1823), which is often credited as initiating the science of thermodynamics.

Sadi Carnot drew inspiration from the design of steam engines. He believed that France had been defeated in the Napoleonic wars because it hadn't harnessed the power of steam to build heavy industry in the way that England had. However, although the steam engine had been invented and successfully commercialized in England, its design had been mostly down to trial and error and the intuition of engineers such as the Scottish inventor James Watt. What it lacked was any theoretical foundation. Carnot sought to rectify this situation by describing in mathematical terms how any heat engine, such as those that drove steam trains, can be used to do work via a cyclical process that is to this day known as the *Carnot cycle*.

The Carnot cycle describes how a heat engine transfers energy from a hot to a cold place and harnesses some of this energy to do useful work, before returning to its initial state. For example, a steam engine transfers heat from the hot boiler to the condenser, where it is cooled, and in the process harnesses some of the heat energy as steam to do the work of moving a piston and thereby the wheels of a locomotive. The cooled water is then returned to the boiler ready to be heated up again for another round of the Carnot cycle.

The principle of the Carnot cycle applies to all kinds of engines that use heat to do any sort of work, from the steam engines that powered the industrial revolution to the petrol engine that drives your car or the electrical pump that cools your fridge. Carnot showed that the efficiency of each of these – in fact, 'every imaginable heat engine', as he put it, depends on a few fundamental principles. Moreover, he proved that the efficiency of any classical heat engine cannot exceed a theoretical maximum, now known as the *Carnot limit*. For example, an electric motor that is using 100 watts of electric power to supply 25 watts of mechanical power has an efficiency of 25 per cent: it is losing 75 per cent of its supplied energy as heat. Classical heat engines are not very efficient.

The principles and limitations of Carnot heat engines are extraordinarily broad and can even be applied to photocells, such as those found on the roofs of some buildings, that capture light energy and convert it into electricity. The same is true of the biological photocells in the chloroplasts of leaves that we have described in this book. Such a *quantum heat engine* does a similar job to a classical heat engine, but with electrons instead of steam and photons of light replacing the heat source. Electrons first absorb photons and are excited to a higher energy. They can then give up this energy, when required, to do useful chemical work. This idea goes back to the work of Albert Einstein and would later underpin the principles of the laser. The problem is that many of these electrons will lose their energy as wasted heat before they have a chance to put it to use. This puts a limit on the efficiency of such a quantum heat engine.

You will remember that the reaction centre is the final destination for all those oscillating excitons in photosynthetic complexes. So far, we have focused on the energy delivery process; but the real action of photosynthesis takes place in the reaction centre itself. Here the fragile energy of excitons is converted into the stable chemical energy of the electron carrier molecule that plants or microbes use to do lots of useful work, like building more plants and microbes.

What takes place in the reaction centre is just as remarkable as the exciton transport step, and even more mysterious. Oxidation is the chemical process by which electrons are moved between atoms. In many oxidations, electrons actively hop from one atom (which becomes oxidized) to another. But in other oxidations, such as the burning of coal, wood or any carbon-based fuel, electrons that are initially possessed solely by one atom end up being shared with other atoms: a net loss of electrons to the electron donor (just as sharing your chocolate bar involves a net loss of chocolate). So, when carbon is burnt in air, electrons in its outer orbits end up being shared with oxygen to form the molecular bonds of carbon dioxide. In these burning reactions the outer electrons of carbon are only relatively loosely bound, so they are relatively easy to share. But in the photosynthetic reaction centre of a plant or microbe, energy is used to pluck electrons right out of water molecules in which the electrons are far more tightly bound. Essentially, a pair of H_2O molecules is split to produce one O_2 molecule, four positively charged hydrogen ions and four electrons. So, since water molecules *lose* their electrons, the reaction centre is the only natural place where water is oxidized.

In 2011 the American physicist Marlan Scully, currently a professor jointly at Texas A&M University and Princeton, along with his co-workers at several US universities, described a clever way for a hypothetical quantum heat engine to be engineered to exceed the efficiency limits of a standard quantum heat engine.[8] To do this, molecular noise is used to nudge an electron into a superposition of two energy states at the same time. When this electron then absorbs the energy of a photon and is 'excited', it will remain in a superposition of two (now higher) energies at once. Now the probability of the electron falling back to its original state and losing its energy as wasted heat can be reduced, thanks to the quantum coherence of its two energy states – this is similar to the example of the interference pattern produced by the double-slit experiment we described in chapter 4. There, certain positions on the back screen that are available to the atom when just one slit is open become inaccessible owing to destructive interference when both slits are open. Here, the delicate collaboration between molecular noise and quantum coherence tunes a quantum heat engine to reduce inefficient wastage of thermal energy and thereby increase its efficiency beyond the quantum Carnot limit.

But is such delicate tuning at the quantum level possible? You would need to engineer at a subatomic scale both the position and the energies of individual electrons to deliver just the right amount of interference to increase the flow of energy along efficient paths and eliminate flow down the wasteful ones. You would also need to tune the surrounding molecular white noise so that it would nudge the off-beat electrons into the same beat; but not too vigorously, or they would be knocked into different rhythms and coherence would be lost. Is there anywhere in the universe where we might expect to find this finely tuned degree of molecular order capable of exploiting delicate quantum effects in the subatomic world?

Scully's 2011 paper was entirely theoretical. No one has yet built a quantum heat engine that can capture the expected Carnot-busting energy bonus. But in 2013 another paper from the same team pointed out a curious fact regarding photosynthetic reaction centres.[9] They are all equipped, not with a single chlorophyll molecule that might be able to operate a straightforward quantum heat engine, but with a pair of chlorophyll molecules known as a *special pair*.

Although the chlorophyll molecules in the special pair are identical, they are embedded in different environments in the protein scaffold, which makes them vibrate at slightly different frequencies: they are slightly out of tune. In their later paper, Scully and his colleagues pointed out that this structure provides photosynthetic reaction centres with the precise molecular architecture needed for them to work as quantum heat engines. The researchers showed that the chlorophyll's special pair appears to be tuned to exploit quantum interference to inhibit inefficient wasteful energy routes and thereby deliver energy to the acceptor molecule with an efficiency that exceeds the nearly 200-year-old limit discovered by Carnot by a margin of between 18 and 27 per cent. That may not seem a huge amount until we consider the current estimate that world energy consumption will grow by about 56 per cent between 2010 and 2040: then developing a technology capable of boosting energy by a comparable margin seems staggeringly important.

This extraordinary result provides yet another remarkable example of how living organisms rooted in the quantum world seem to have abilities denied to inanimate macroscopic machines. Of course, quantum coherence is necessary for this scenario to work; but in another stop-press result published in July 2014, a team of researchers from the Netherlands, Sweden and Russia detected quantum beating in plant photosystem reaction centre II* and went on to claim that these centres function as 'quantum-designed light traps'.[10] And remember that photosynthetic reaction centres evolved between two and three billion years ago. So for nearly the entire history of our planet, plants and microbes seem to have been utilizing quantum-boosted heat engines – a process so complex and clever that we have yet to work out how to reproduce it artificially – to pump energy into carbon and thereby make all the biomass that formed microbes, plants, dinosaurs and, of course, us. Indeed, we are still harvesting ancient quantum energy in the form of fossil fuels that warm our

---

* Plants have two photosystems, I and II.

homes and power our cars and drive most of today's industry. The potential benefits of modern human technology learning from ancient natural quantum technology are huge.

So, in photosynthesis, noise seems to be utilized both to enhance the efficiency of delivery of excitons to the reaction centre and to capture that solar-derived energy once it arrives in the reaction centre. But this ability to make a quantum virtue out of a molecular vice – noise – is not limited to photosynthesis. In 2013, Nigel Scrutton's group at the University of Manchester, the team who studied proton tunnelling in enzymes in the experiments that we discussed in chapter 3, replaced the regular atoms in an enzyme with heavier isotopes. The isotope exchange had the effect of adding extra weight to the protein's molecular springs so that they vibrated – its coloured noise – at different frequencies. The researchers found that proton tunnelling and enzyme activity were perturbed in the heavier enzyme,[11] suggesting that in the normal state with the natural, lighter isotopes, the metronome-like oscillations of its protein backbone contribute to tunnelling and enzyme activity. Similar results, with other enzymes, have been obtained by Judith Klinman's group at the University of California.[12] So, as well as guiding photosynthesis, noise appears also to be involved in boosting enzyme action. And remember that enzymes are the engines of life that have made every single molecule inside every cell of every living creature on our planet. Good vibrations may be playing a crucial role in keeping us all alive.

Life on the quantum edge of a classical storm

On a ship at sea: a tempestuous noise

William Shakespeare, *The Tempest*,
Act I, scene 1, opening stage direction

Do any of these insights provide an answer to the question Schrödinger posed decades ago about the nature of life? We have already taken on board his insight that life is a system dominated by order that goes all the way down, from highly organized whole organisms through the stormy thermodynamic ocean to the quantum bedrock below (figure 10.1). And, crucially, these dynamics of life are delicately poised and balanced so that quantum-level events can make a difference to the macroscopic world, just as Pascual Jordan predicted in the 1930s. This macroscopic sensitivity to the quantum realm is unique to life and allows it potentially to exploit quantum-level phenomena, such as tunnelling, coherence and entanglement, to make a difference to us all.

But, and this is a big but, this exploitation of the quantum world can only take place if decoherence can be kept at bay. Otherwise the system loses its quantum character and behaves entirely classically or thermodynamically, relying on the 'order from disorder' rules. Scientists have fended off decoherence by shielding their quantum reactions from intrusive 'noise'. This chapter has revealed that life appears to have adopted a very different strategy. Instead of allowing noise to hinder coherence, life uses noise to maintain its connection to the quantum realm. In chapter 6, we imagined life as a granite block delicately poised to render it susceptible to quantum-level events. For reasons that will become clear in a moment, we will make a metaphorical shift by replacing our granite block with a tall sailing ship.

Our imaginary sailing ship will initially be in dry dock, with its narrow keel exquisitely balanced on a single line of carefully aligned atoms. In this perilously poised state our ship, like a living cell, is sensitive to quantum-level events taking place in its atomic keel. The tunnelling of a proton, the excitation of an electron or the entanglement of an atom can all have an influence on the entire ship, perhaps by affecting its delicate balance on the dry dock. However, we will further imagine that its captain has found clever and surprising ways to make good use of these delicate quantum phenomena such as coherence, tunnelling, superposition or entanglement to help navigate his craft once it sets sail.

But remember that we are still in dry dock: this ship isn't going anywhere just yet. And although in its delicately balanced state it can potentially harness quantum-level phenomena, its precarious perch leaves it vulnerable to even the faintest imaginable breeze – perhaps being touched by just a single air molecule – which could topple the whole vessel. The engineer's approach to the problem of keeping the craft upright and thus retaining its sensitivity to the quantum events in its keel would be to enclose the ship in a shielded box and pump out all the air to prevent any stray billiard-ball-like molecule from disturbing the vessel. The engineer would also cool the entire system down to close to absolute zero, so that not even a molecular vibration could disturb its delicate balance. But skilled sea captains know that there is another way to keep a ship upright: it must first be launched into turbulent thermodynamic waters.

We take it for granted that a ship is easier to keep upright in water than on land, but thinking about it at a molecular level we find that the reason for its increased stability isn't immediately obvious. We have just said that an engineer's approach to keeping a narrow-keeled ship upright in dry dock would be to protect the vessel from any potential disturbance from stray atoms or molecules. But isn't the sea full of stray atoms and molecules randomly jostling each other and the keel of any ship in that billiard-ball fashion that we explored in chapter 2? How is it that the precariously balanced ship can be toppled by tiny impacts on land but remains impervious to them when on the water?

The answer comes back to those 'order from disorder' rules that Schrödinger described. The ship will indeed be bombarded with trillions of molecular impacts from both its port and starboard sides. Of course, it is now no longer balancing on its ultra-thin keel but kept afloat by the buoyancy of the water, and with so many impacts on both sides of the ship, the average force to the bow and stern or to port and starboard will be the same. So buoyant ships do not topple because they are being held up by trillions of random molecular bombardments: order (the ship's vertical orientation) from disorder (trillions of random billiard-ball-like molecular impacts).

But ships can of course be toppled, even on the high seas. Imagine that the captain has launched his ship onto a tempestuous sea, but hasn't yet hoisted its sail. The waves buffeting the vessel aren't so random any more, and big swells may surge from one side or another that could easily topple an unstable vessel. But our clever captain knows how to increase the stability of his ship: he hoists the sails so that he can harness the power of the wind to keep his vessel on an even keel (Figure 10.2).

Once again, this stratagem may, at first sight, appear to be contradictory. We would expect that haphazard winds and unpredictable gusts would act to topple rather than stabilize an already unsteady ship – particularly as they won't be random but will tend to arrive with more force on one side of the ship or the other. But the captain knows how to adjust the angle of the sail and the tiller so that the action of the wind and currents acts against the gusts and the gales to

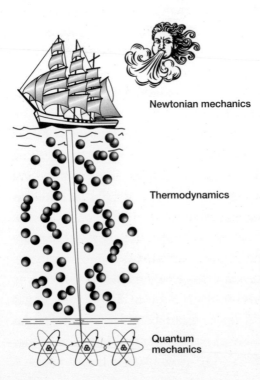

Newtonian mechanics

Thermodynamics

Quantum mechanics

Figure 10.2: Life navigates the edge of the quantum and classical worlds. The living cell is like a ship whose narrow keel penetrates right down to the quantum layer of reality and can thereby capture phenomena such as tunnelling or entanglement to keep itself alive. This connection to the quantum realm has to be actively maintained by living cells harnessing the thermodynamic storms – molecular noise – to maintain, rather than disrupt, quantum coherence.

correct any listing to one side or the other. In this way, he can harness the surrounding tempest to keep his ship stable.

Life, it seems, is like that metaphorical ship sailing through stormy classical waters with a clever captain on board: the genetic programme, honed by nearly four billion years of evolution, is able to navigate the various depths of the quantum and classical realms. Rather than hiding from the tempests, life embraces them, marshalling their molecular squalls and gales to fill its sails and keep the ship upright so that its narrow keel penetrates the thermodynamic waters to connect with the quantum world (figure 10.2). Life's deep roots allow it to harness those weird phenomena that prowl the quantum edge.

Does this provide us with a new insight into what life really is? Well, there is one further speculation, and we emphasize that it is indeed speculation, but one that, having journeyed this far, we cannot resist making. Remember the question that we posed in chapter 2 concerning the difference between the living and the lifeless; that difference that the ancients described as our soul? Death, they believed, was brought about by the departure of the soul from the body. Descartes' mechanistic philosophy expelled vitalism and discarded the soul, at least from plants and animals, but the difference between the living and the dead remained mysterious. Can our new understanding of life replace the soul with a quantum vital spark? Many will regard the very posing of this question as suspect, pushing the boundaries of conventional science beyond respectability and into the realms of pseudoscience or even a kind of spirituality. That is not what we are proposing here. Instead, we want to offer what we hope is an idea that might replace mystical and metaphysical speculations with at least the grain of a scientific theory.

In chapter 2, we compared the capacity of life to preserve its highly organized state to a billiard-table contraption that could maintain a triangle of balls in the centre of the table by detecting and replacing any balls knocked out of place by the collisions from other balls in a thermodynamic-like system. Now that you have discovered more about how life works, you can see that this self-sustainability is maintained by the complex molecular machinery of enzymes, pigments, DNA, RNA and other biomolecules, some of whose properties depend on quantum mechanical phenomena such as tunnelling, coherence and entanglement.

The recent evidence that we have examined in this chapter suggests that some or all of these diverse quantum-boosted activities, which we could imagine as the activities taking place on the busy deck of our ship, are maintained by life's remarkable ability to harness thermodynamic storms and gales to retain its connection with the deeper quantum realm. But what happens if the thermodynamic storm blows too strongly, metaphorically breaking the ship's mast? No longer able to harness the thermodynamic gusts and gales – the white and coloured noise – to maintain an even keel, the sail-less cell will be buffeted by the waves and swells of its interior, causing our metaphorical ship to pitch and roll, eventually severing its

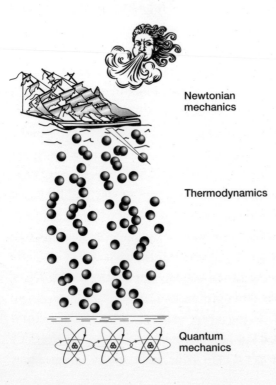

Newtonian
mechanics

Thermodynamics

Quantum
mechanics

Figure 10.3: Perhaps death represents the severing of the living organism's connection with the orderly quantum realm, leaving it powerless to resist the randomizing forces of thermodynamics.

connection with the orderly quantum realm (figure 10.3). Without this connection, coherence, entanglement, tunnelling or superposition can no longer influence the cell's macroscopic behaviour, so the quantum-disconnected cell will sink beneath the thermodynamically turbulent waters, becoming an entirely classical object. Once a ship has sunk, no storm will refloat it; and perhaps, once a living organism has been captured by the stormy ocean of molecular motion, no tempest can restore its quantum connection.

Can we exploit quantum biology to make new living technology?

Storms may not be able to refloat a sunken boat, but humans can. Human ingenuity can achieve so much more than random forces. As we discussed in chapter 9, the probability of a tornado blowing mindlessly through a junkyard and assembling a jumbo jet through sheer chance is stupendously tiny. But aviation engineers can build planes. Can we also assemble life? As we have pointed out on several occasions in this book, no one has yet succeeded in making life out of inert chemicals; which, according to Richard Feynman's famous dictum, means that we do not yet fully understand the phenomenon of life. But perhaps our new-found understanding of quantum biology can provide us with the means to create new life and even build a revolutionary form of *living technology*.

Living technology is of course familiar. We are completely dependent on it, in the form of agriculture, to make our food. We also rely on the products of living technology, such as the bread, cheese, beers and wines that have been transformed from flour, milk, grain and fruit juices by yeast and bacteria. Our modern world similarly benefits from its harvest of the non-living products of once-living cells, such as the enzymes that Mary Schweitzer used to break down dinosaur bone. Similar enzymes are used to break down natural fibres to make the fabrics for clothes; or are added to the biological detergents that wash those clothes. The multimillion-dollar biotechnology and pharmacology industries produce hundreds of natural products, such as the antibiotics that protect us from infection. The energy industry exploits the ability of microbes to turn excess biomass into biofuels; and many of the materials that support modern life, such as timber and paper, were once alive, as were the fossil fuels that heat our homes and power our cars. So even in the twenty-first century we remain extraordinarily reliant on our millennia-old living technology. If you harbour any remaining doubts, try reading Cormac McCarthy's *The Road*, a dystopian novel describing the bleak world that would be left to us if we carelessly destroyed our living technology.

But the existing living technology has its limitations. For example, although – as we have discovered – some of the steps in the process of photosynthesis are extraordinarily efficient, most are not, and the overall energy efficiency of the conversion of solar energy into chemical energy that we can harvest in agriculture is very low. The reason is that plants and microbes have a different agenda from ours: they carry out inefficient chores like making flowers and seeds that are not really necessary for energy capture, but are nevertheless essential for their own survival. Similarly, the microbes that make antibiotics, enzymes or pharmaceuticals do so very wastefully because their evolution-honed agenda forces them to make lots of unnecessary stuff, like more microbial cells.

Can we engineer life that sticks to our agenda? Of course we can, and we already benefit hugely from humankind's successful transformation of wild plants and animals into the living technology of domesticated crops and livestock, optimized for human exploitation. But the process of artificial selection that gave us plants with bigger seeds or docile animals suitable for farming, although highly successful, has its limitations. We cannot select what nature has not already fashioned. For example, billions of dollars are spent each year on fertilizers that replenish the soil's nitrogen, lost through intensive agriculture. Leguminous crops, such as peas, don't need nitrogen fertilizers because they harbour bacteria in their roots that capture the gas directly from the air. Agriculture could be far more efficient if we could engineer leguminous cereal crops that fix their own nitrogen, like peas do. But this capability hasn't evolved in any cereal.

However, even this limitation can be at least partly overcome. The genetic manipulation of plants, microbes and even animals (genetic engineering) took off in the late twentieth century. Today, much of the harvest of major crops such as soya comes from genetically engineered plants resistant to disease or herbicide, and efforts are under way to, for example, insert nitrogen-capturing genes into cereal crops. The biotechnology industry similarly relies heavily on genetically modified microbes to produce our pharmaceuticals and antibiotics.

Even so, once again there are limitations. Genetic engineering mostly just moves genes around from one species to another. For example, the rice plant makes vitamin A (beta carotene) in its leaves but not in its seed, so hardly any is present in the staple crop that feeds much of the developing world. Vitamin A is essential for our immune system and our vision, so its deficiency in the poorest rice-dependent regions of the world leads to millions of children dying of infections or becoming blind each year. In the 1990s Peter Beyer from the University of Freiberg and Ingo Potrykus from the Swiss Federal Institute of Technology in Zurich genetically inserted two genes needed to make vitamin A, one from daffodils and the other from a microbe, into the rice genome to make rice with high levels of vitamin A in its seeds. *Golden rice*, as Beyer called it because of the yellow colour of its seeds, can now provide most of the daily vitamin A requirement for children. However, although genetic engineering is a very successful technology, it is really just tinkering with life. The new science of *synthetic biology* aims to make a truly revolutionary living technology by engineering entirely novel forms of life.

There are two complementary approaches to synthetic biology. The top-down approach is the one we have already met in the course of discussing how the genome-sequencing pioneer Craig Venter constructed so-called 'synthetic life' by replacing the genome of a bacterium called mycoplasma with a chemically synthesized version of the same genome. This genome swap allowed his team to make relatively minor modifications to the entire mycoplasma genome. Nevertheless, it was still a mycoplasma: they did not introduce any radical changes into the bacterium's biology. Over the coming years Venter's team plan to engineer more radical changes; but these changes will be introduced step-by-step in this top-down synthetic biology approach. The team did not make new life: they modified existing life.

The second approach is bottom-up and is far more radical: rather than modifying an existing living organism, bottom-up synthetic biology aims to engineer completely new life forms out of inert chemicals. Many would consider such an endeavour dangerous, even sacrilegious. Is it even feasible? Well, living organisms, like us, are extraordinarily elaborate machines. Like any machine, they can be reverse-engineered to discover their design principles; and those design principles can then be harnessed to build even better machines.

Building life from the bottom up

Bottom-up synthetic life enthusiasts dream of making completely novel life forms that could transform our world. For example, today's architects are rightly preoccupied with the notion of sustainability: sustainable houses, offices, factories and cities. However, although modern buildings or cities are often described as self-sustainable, they mostly rely on the efforts and skills of truly self-sustainable beings, humans, to keep them in shape: when your roof tiles are blown off in a storm, you hire the builder to climb up and replace them; when your pipes leak, you call the plumber; when your car breaks down, you have it towed to a mechanic. Essentially, all this manual maintenance is needed to repair the damage inflicted on our homes or machines by all that billiard-ball-like molecular jostling inflicted by wind, rain and other environmental insults.

Life is different: our bodies are able to continually maintain themselves by renewing, replacing and repairing damaged or worn-out tissue. For as long as we are alive, we are indeed self-sustainable. Modern architecture has attempted to mimic certain features of life in many of the signature buildings of recent years. For example, Norman Foster's 'Gherkin' tower, which was added to the London skyline in 2003, possesses a hexagonal skin inspired by the Venus Flower Basket Sponge (which is a sponge, not a flower) that efficiently distributes the building's stresses. The Eastgate Centre in Harare, Zimbabwe, designed by architect Mick Pearce, mimics the air-conditioning system of termite mounds to provide ventilation and cooling. Rachel Armstrong, co-director of the architectural research group AVATAR based at the University of Greenwich, has a bolder vision: truly self-sustaining buildings, the ultimate biometric architecture. Along with several other architectural visionaries, she dreams of engineering buildings out of artificial living cells that possess the ability to sustain, self-repair and even self-replicate.[13] If such living buildings were damaged by wind, rain or flood they would, like living organisms, sense their injury and self-repair just like living bodies.

Armstrong's ideas could be extended to enhance other synthetic features of our lives. Living material could also be used to construct prosthetics, such as artificial limbs or joints, that would be able to self-repair and protect themselves from microbial attack, just like living tissue. Artificial life forms could even be injected into the human body to, for example, search out and destroy cancer cells. Pharmaceuticals, fuels and food could all be made by custom-engineered synthetic life forms unencumbered by any evolutionary history. Further into the future is the sci-fi vision of living robots, androids, which could take over the menial tasks of society or even 'terraform' Mars to make it habitable for human colonies, or construct living spaceships that could explore the galaxy.

The idea of the bottom-up creation of synthetic life can be traced back to the early twentieth century, when the French biologist Stéphane Ludec wrote that, 'just as synthetic chemistry began with artificial formation of the simplest organic products, so biological

321

synthesis must content itself at first with the fabrication of forms resembling those of the lowest organisms'.[14] As we discussed in chapter 9, even the 'lowest organisms' alive today are actually extraordinarily complex bacteria made up of thousands of parts that cannot currently be synthesized by any conceivable bottom-up approach. Life must have started from something much simpler than a bacterium. Today's best guesses for our ultimate ancestor are, as we suggested in that chapter, molecules of self-replicating enzymatic RNA (ribosomes) or protein that became enclosed within some kind of small vesicle to form a self-replicating simple cellular structure, a *protocell*. The nature of the first protocells, if they indeed existed, is not at all clear. Many scientists believe they sheltered within microscopic pores in rocks, such as the Isua rocks that we met in chapter 9, filled with simple biochemicals capable of supporting life. Others believe that they were bubbles or droplets of biochemicals bounded by some kind of membrane floating in the primordial ocean.

Most bottom-up synthetic life enthusiasts take inspiration from origin-of-life theories by attempting to build their own artificial living protocells capable of swimming through a laboratory-based primordial sea. Probably the simplest are the various kinds of droplets or vesicles of oil in water or water in oil. These are easy to make; in fact, you have made millions of them whenever you made a salad dressing. Oil and water famously don't combine and so will quickly separate; but if you add a substance whose molecules embed between water and oil, a *surfactant* such as mustard, and give the mixture a good shake, you make a salad dressing. Although this may look smooth and homogeneous, in reality it is filled with trillions of tiny stable oil droplets.

Martin Hanczyc of the University of Southern Denmark has made remarkably life-like protocells out of oil-in-water droplets that are stabilized by detergent. His protocells are very simple, often constructed from only five chemicals. Mixed in the right proportions, they self-assemble into oily droplets. The interior of the droplets supports a simple chemistry that causes the protocell to move through its environment, propelled by convection (heat circulation) and the same kind of chemical forces that cause oil droplets to coalesce in the first place. They are even able to undergo a simple form of growth and self-replication by absorbing raw materials from their environment, eventually causing them to break in two.[15]

Hanczyc's protocells are inside-out compared to living cells, as they have an oily interior and water on the outside. Most other researchers opt for making protocells with a watery interior. This also allows them to fill them with ready-made water-soluble bio-molecules. For example, in 2005 the geneticist Jack Szostak filled protocells with RNA ribozymes.[16] Remember (chapter 9) that ribozymes are RNA molecules that can encode genetic information, just like DNA, but also host enzyme activity. The team showed that the ribozyme-filled protocells were capable of a simple form of her-edity, essentially splitting in two like Hanczyc's protocell. In 2014 a team led by Sebastien Lecommandoux based at Radboud University in the Netherlands made another kind of protocell whose multiple compartments were filled with enzymes that could, like living cells, support a simple metabolism that cascaded from one compartment to another.[17]

These dynamic, chemically active protocells are certainly intriguing and impressive constructions; but are they life? To answer this question we need to agree on a working definition of life. The obvious one, self-replication, is fine for many purposes, but it asks too much. Most of the cells in an adult body, such as red blood cells and nerve cells, do not replicate, yet they are undoubtedly alive. Even whole humans, such as Buddhist or Catholic priests, don't (usually) bother with the messy business of self-replication, yet they remain very much alive. So, although self-replication is of course necessary for the long-term survival of any species, it is not an obligatory property of life.

A property of life that is even more fundamental than self-replication is the one that we have already discussed and which biomimetic architects strive to emulate: self-sustainability. Life is able to sustain its living state. So the minimum requirement we will demand for our bottom-up protocells to earn classification as living is that they must be capable of sustaining themselves in turbulent thermodynamic seas.

Unfortunately, using this more limited definition of life, none of the existing generation of protocells is alive. Even those that can perform a few tricks, like a simple form of replication (splitting in two), produce daughters that are not really the same as the parents: they have less of the starting components, such as ribozymes or enzymes, so that, as the replication process proceeds, these components are eventually exhausted. Similarly, although protocells such as those made by Lecommandoux's group are able to support a metabolism resembling that of a simple living organism, they need to be filled with active biomolecules, which they are not capable of replenishing themselves. The current generation of protocells are like wound-up clocks: they can maintain their chemical start-up state, supported by premade enzymes and substrates, until it runs down. Thereafter, the continual battering received from the surrounding molecular motion erodes the organization of these protocells so that they become progressively more chaotic and random until they are eventually no different from their environment. Artificial protocells are, unlike life, incapable of winding themselves up.

Are they missing an ingredient? The field is of course very young, and it is likely that great strides will be made in the coming decades. The idea we want to explore in this last section of our book is that quantum mechanics could provide the missing spark needed to animate artificial life and make truly synthetic life. As well as launching a revolutionary technology, such an advance might also finally provide us with the means to answer that ancient question that we posed in chapter 2: what is life?

We, and others, have argued that the thermodynamic description of life is inadequate as it does not incorporate life's ability to harness the quantum realm. Life, we believe, depends on quantum mechanics. But are we right? As we have already discussed, this is hard to prove with the technology we have today, because you can't just turn quantum mechanics off and on in a living cell. However, we predict that life, whether natural or artificial, is impossible without the strange features of the quantum world we have discussed in this book. The only way to find out if we are right is to make synthetic life with and (if possible) without quantum weirdness and see which one works best.

Launching the primordial quantum protocell

Let us imagine building a simple living cell out of totally inanimate material; perhaps one able to perform simple tasks, such as finding its own food within a kind of laboratory-maintained primordial sea. Our aim will be to build such a device in two ways. One will seek to harness the weird features of quantum mechanics – we will call it the *quantum protocell*. The other will not – this will be the *classical protocell*.

A good starting point for both versions would be Sebastien Lecommandoux's multi-compartment, membrane-bounded proto-cells, the different sections of which allow us to separate the distinct functions of life into individual compartments. Next, we need to provide our protocell vessel with an energy source: let's use that abundant source of high-energy photons, sunlight. We will load one of its compartments with a forest of pigment molecules and scaffold proteins, making a form of solar panel, able to capture photons and convert their energy into excitons, like an artificial chloroplast. However, jumbled-up pigment molecules will be unlikely to deliver the high-efficiency energy transport characteristic of photosynthesis, since the molecular muddle will be unable to maintain the quantum coherence needed for efficient energy transport. To capture the quantum beat we need to orientate the pigment molecules so that the coherent wave can flow through the system.

In 2013, a University of Chicago group led by quantum photosynthesis pioneer Greg Engel tackled this problem by chemically bolting pigment molecules together in a fixed alignment. Just like the algal FMO complex in which Engel had first detected quantum coherence (chapter 4), their artificial pigment system showed coherent quantum beats that continued for tens of femtoseconds, even at room temperature.[18] So, to provide the solar panel of our quantum protocell with coherence-boosted excitons, we will fill it with a forest of Engel's bolted-together pigment molecules. The classical photocell will contain the same pigments, but they will be randomly aligned so that the exciton will have to meander its way through the system. We could thereby test whether quantum coherence is essential or dispensable for exciton transport in photosynthesis.

However, as we have discovered, catching light is only the first step in the job of photosynthesis; we next need to transform the unstable exciton energy into a stable chemical form. Once again, some progress has already been made. When the Scully group demonstrated in their 2013 paper that the photosynthetic reaction centre appears to be a quantum heat engine, they went on to argue that biological quantum heat engines could inspire the design of more efficient photocells.[19] Later that same year, a team from the University of Cambridge took them at their word and produced a detailed blueprint for such an artificial photocell that would work as a quantum heat engine.[20] The group modelled an artificial reaction centre from the bolted-together pigment molecule made in Engel's laboratory and showed that it should deliver an energetic electron to an acceptor molecule with a similar Carnot limit-busting efficiency enhancement as the Scully group had discovered for natural photosynthesis.

So let us imagine our quantum solar cell rigged onto an artificial reaction centre, inspired by the Cambridge team's model, that is able to capture energetic electrons as stable chemical energy. Once again, we will engineer a rival system for our classical protocell that attempts a similar energy transfer process, but without the quantum Carnot-busting efficiency. Once the light energy has been captured, it can be used to build complex biomolecules, such as the cell's pigment molecules.

However, as well as electrons, biosynthetic reactions need an additional energy boost that, in our own cells, is provided by cellular respiration (chapter 3). We will take inspiration from respiration and shunt some of the high-energy electrons delivered by photosynthesis into a 'power plant' compartment, where they will tunnel from one enzyme to another, as in natural respiratory chains, to make ATP, the cell's molecular energy carrier. Once again, our aim will be to engineer the respiratory compartment and to explore the role of quantum mechanics in this vital biological process.

With a source of electrons and energy, our quantum protocell is now equipped to make all its own biochemicals; but it needs a source of raw material – food. So we provide it with a food source, a simple sugar: glucose dissolved in our laboratory-based primordial sea. We will have to install ATP-powered sugar transporters, able to pump the glucose inside the cell, together with another suite of enzymes, capable of manipulating its atoms – quantum-level engineering – to build more complex biomolecules. Many of these enzymes normally utilize electron and proton tunnelling, as we discussed in chapter 3, but our aim will be to engineer versions that work with and without the capability of dipping into the quantum world to discover whether quantum mechanics really provides an essential lubricant for these engines of life.

Another feature that we would like to engineer into our quantum-supported protocell is the capability of harnessing the tempest of molecular noise to maintain quantum coherence. At present, too little is known about how life manages this trick to have any confidence in how it might be engineered. Many factors may be involved: for example, the extremely crowded molecular environment of living cells is known to modify many biochemical reactions,[21] and might help to restrict the randomizing impact of noise. So we will pack the protocells very tightly with biomolecules to simulate that crowded living environment in the hope that it will help to harness those thermodynamic squalls and gales to maintain quantum coherence.

But our quantum protocell remains a very needy vessel, since all its enzymes have to be preloaded on board. To make it self-sufficient we must furnish another compartment, its control room, with an artificial DNA-based genome able to encode everything it needs, together with the machinery needed to turn its quantum-level proton code into proteins. This is similar to the top-down approach utilized by Craig Venter; only our genome will be injected into a *non-living* protocell. Lastly, we could even endow our protocell with a navigational system, perhaps a molecular nose to enable it to locate its food by utilizing the quantum entanglement olfactory receptor principle that we explored in chapter 5 and a molecular motor to propel itself through its primordial sea. We could even equip it with a quantum-powered navigational system, like our robin's, which could help it to orientate itself in the laboratory-based primordial ocean.

What we have described is little more than a biological whimsy – no more real than Shakespeare's Ariel. We have omitted a huge amount of detail and, in the interests of simplicity and intelligibility, failed to mention the colossal challenges that would face any real bottom-up synthetic biology project. Even if such a project were ever attempted, it would certainly not try to instantiate all these processes in a single step, as in our imaginary recipe above, but would instead first attempt to install the simplest or the best-understood process – maybe photosynthesis – into a protocell. This would, of course, be a major achievement in itself, and would be the perfect model system to use to investigate the role of quantum coherence in photosynthesis. Were such a feat indeed proved possible, the next steps would be to include additional components to implement greater and greater complexity, leading eventually, perhaps, to a truly artificial living cell. But this will only be possible, we predict, along the quantum route to life: life simply won't work, we believe, without being connected to the quantum realm.

If such a project were indeed ever undertaken, then it might be possible, at last, to make new life. Such an advance could launch a truly revolutionary living technology: artificial life able to navigate the edge between the quantum and classical worlds. Artificial living cells could be engineered to serve as the bricks of truly sustainable living buildings; micro-surgeons could be constructed to repair and replace our damaged and worn tissues. The fantastic features of quantum biology that we have explored in this book, from photosynthesis to enzyme action, and from quantum noses to quantum genomes, quantum compasses and maybe even quantum brains, could all be harvested to potentially build a brave new world of quantum synthetic living organisms that could free their natural-born relatives from the drudgery of providing humanity with most of its needs.

But, perhaps even more importantly, the ability to make new life from scratch would finally provide biology with a reply to answer Feynman's famous dictum that 'what I can't make, I don't understand'. If such a project were indeed successful, then we could, finally, claim that we do at last understand life and its remarkable ability to harness the forces of chaos to sail that narrow edge between the classical and quantum worlds.

> The noontide sun, call'd forth the mutinous winds,
> And 'twixt the green sea and the azured vault
> Set roaring war: to the dread rattling thunder . . .
>   William Shakespeare,
>   *The Tempest*, Act V, scene 1

Epilogue: quantum life

THE EUROPEAN robin we met in chapter 1 has successfully overwintered in the Mediterranean sunshine and is now hopping between the sparse woodland and ancient stones of Carthage in Tunisia, fattening herself on flies, beetles, worms and seeds, all composed of biomass spun out of air and light by the quantum-powered photosynthetic machines we call plants and microbes. But the sun now climbs high in the midday sky and its fierce heat has dried the shallow streams that wind through the woodland. The forest is becoming parched and inhospitable to our European passerine. It is time for her to move on.

It is now late in the day, and the tiny bird flies up to perch on a branch high in a cedar tree. She carefully preens herself, just as she did many months before, while listening to the calls of other robins who have similarly felt an avian urge to ready themselves for a long flight. As the last rays of the sun dip below the horizon, the robin turns her beak towards the north, spreads her wings and launches herself into the evening sky.

The robin flies towards the North African coast and continues across the Mediterranean, taking pretty much the same route, but in the opposite direction, as she took six months earlier, guided once again by her avian compass with its quantum entangled needle. Every beat of her wings is powered by contraction of muscle fibres whose energy has been delivered by quantum tunnelling of electrons and protons through respiratory enzymes. After many hours she reaches the coast of Spain and alights within a forested river valley of Andalucía, where she rests surrounded by abundant vegetation including willows, maple, elm and alder, fruit trees and flowering shrubs such as oleander, each a product of quantum-powered photosynthesis. Odorant molecules waft into her nasal passages, locking onto odour receptor molecules and triggering quantum tunnelling events that send nerve signals, via quantum coherent ion channels, to her brain telling her that citrus flowers are nearby, attended by tasty bees and other pollinating insects that will provide her with additional sustenance for the next stage of her journey.

After many days of flight the robin finally finds her way back to the Scandinavian spruce forest from which she set out many months before. Her first job is to search out a mate. Male robins arrived several days earlier and most have found suitable nesting sites which they advertise to the females with their song. Our robin is attracted to a particularly tuneful bird and, as part of their courtship ritual, enjoys several tasty grubs collected by the male. After a brief coupling the male's sperm is united with the female's egg cell and the quantum-based genetic information encoding the form, structure, biochemistry, physiology, anatomy and even song of each pair of birds is almost flawlessly copied into a new generation of robins. The few quantum-tunnelled errors will provide the raw material for future evolution of the species.

Of course, as we have emphasized in previous chapters, we cannot yet be sure that all the features we have just described are quantum mechanical. But there is no doubt that much of what is or was wonderful and unique about robins, clownfish, bacteria that survive beneath the Antarctic ice, dinosaurs that roamed the Jurassic forests, monarch butterflies, fruit flies, plants and microbes derives from the fact that, like us, they are rooted in the quantum world. There is much that remains to be discovered; but the beauty of any new area of research is the sheer unknown. As Isaac Newton said:

> I do not know what I may appear to the world, but to myself I seem to have been only like a boy playing on the sea-shore, and diverting myself in now and then finding a smoother pebble or a prettier shell than ordinary, while the great ocean of truth lay all undiscovered before me.

Notes

Chapter 1: Introduction

1 P. W. Atkins, 'Magnetic field effects', *Chemistry in Britain*, vol. 12 (1976), p. 214.

2 S. Emlen, W. Wiltschko, N. Demong and R. Wiltschko, 'Magnetic direction finding: evidence for its use in migratory indigo buntings', *Science*, vol. 193 (1976), pp. 505–8.

Chapter 2: What is life?

1 S. Harris, 'Chemical potential: turning carbon dioxide into fuel', *The Engineer*, 9 August 2012, http://www.theengineer.co.uk/energy-and-environment/in-depth/chemical-potential-turning-carbon-dioxide-into-fuel/1013459.article#ixzz2upriFA00.

2 *Die Naturwissenschaften*, vol. 20 (1932), pp. 815–21.

3 Pascual Jordan, 1938, quoted in P. Galison, M. Gordin and D. Kaiser, eds, *Quantum Mechanics: Science and Society* (London: Routledge, 2002), p. 346.

4 H. C. Longuet-Higgins, 'Quantum mechanics and biology', *Biophysical Journal*, vol. 2 (1962), pp. 207–15.

5 M. P. Murphy and L. A. J. O' Neil, eds, *What is Life? The Next Fifty Years: Speculations on the Future of Biology* (Cambridge: Cambridge University Press, 1995).

Chapter 3: The engines of life

1 R. P. Feynman, R. B. Leighton and M. L. Sands, *The Feynman Lectures on Physics* (Reading, MA: Addison-Wesley, 1964), vol. 1, pp. 3–6.

2 M. H. Schweitzer, Z. Suo, R. Avci, J. M. Asara, M. A. Allen, F. T. Arce and J. R. Horner, 'Analyses of soft tissue from Tyrannosaurus rex suggest the presence of protein', *Science*, vol. 316: 5822 (2007), pp. 277–80.

3 J. Gross, 'How tadpoles lose their tails: path to discovery of the first matrix metalloproteinase', *Matrix Biology*, vol. 23: 1 (2004), pp. 3–13.

4 G. E. Lienhard, 'Enzymatic catalysis and transition-state theory', *Science*, vol. 180: 4082 (1973), pp. 149–54.

5 C. Tallant, A. Marrero and F. X. Gomis-Ruth, 'Matrix metalloproteinases: fold and function of their catalytic domains', *Biochimica et Biophysica Acta (Molecular Cell Research)*, vol. 1803: 1 (2010), pp. 20–8.

6 A. J. Kirby, 'The potential of catalytic antibodies', *Acta Chemica Scandinavica*, vol. 50: 3 (1996), pp. 203–10.

7 Don DeVault and Britton Chance, 'Studies of photosynthesis using a pulsed laser: I. Temperature dependence of cytochrome oxidation rate in chromatium. Evidence for tunneling', *BioPhysics*, vol. 6 (1966), p. 825.

8 J. J. Hopfield, 'Electron transfer between biological molecules by thermally activated tunneling', *Proceedings of the National Academy of Sciences*, vol. 71 (1974), pp. 3640–4.

9 Yuan Cha, Christopher J. Murray and Judith Klinman, 'Hydrogen tunnelling in enzyme reactions', *Science*, vol. 243: 3896 (1989), pp. 1325–30.

10 L. Masgrau, J. Basran, P. Hothi, M. J. Sutcliffe and N. S. Scrutton, 'Hydrogen tunneling in quinoproteins', *Archives of Biochemistry and Biophysics*, vol. 428: 1 (2004), pp. 41–51; L. Masgrau, A. Roujeinikova, L. O. Johannissen, P. Hothi, J. Basran, K. E. Ranaghan, A. J. Mulholland, M. J. Sutcliffe, N. S. Scrutton and D. Leys, 'Atomic description of an enzyme reaction dominated by proton tunneling', *Science*, vol. 312: 5771 (2006), pp. 237–41.

11 David R. Glowacki, Jeremy N. Harvey and Adrian J. Mulholland, 'Taking Ockham's razor to enzyme dynamics and catalysis', *Nature Chemistry*, vol. 4 (2012), pp. 169–76.

Chapter 4: The quantum beat

1 From the BBC TV series *Fun to Imagine 2: Fire* (1983), available on YouTube: http://www.youtube.com/watch?v=ITpDrdtGAmo.

2 Interview with CBC News, available at: http://www.cbc.ca/news/technology/quantum-weirdness-used-by-plants-animals-1.912061.

3 G. S. Engel, T. R. Calhoun, E. L. Read, T-K. Ahn, T. Mančal, Y-C. Cheng, R. E. Blankenship and G. R. Fleming, 'Evidence for wavelike energy transfer through quantum coherence in photosynthetic systems', *Nature*, vol. 446 (2007), pp. 782–6.

4 I. P. Mercer, Y. C. El-Taha, N. Kajumba, J. P. Marangos, J. W. G. Tisch, M. Gabrielsen, R. J. Cogdell, E. Springate and E. Turcu, 'Instantaneous mapping of coherently coupled electronic transitions and energy transfers in a photosynthetic complex using angle-resolved coherent

optical wave-mixing', *Physical Review Letters*, vol. 102: 5 (2009), pp. 057402.

5   E. Collini, C. Y. Wong, K. E. Wilk, P. M. Curmi, P. Brumer and G. D. Scholes, 'Coherently wired light-harvesting in photosynthetic marine algae at ambient temperature', *Nature*, vol. 463: 7281 (2010), pp. 644–7.

6   G. Panitchayangkoon, D. Hayes, K. A. Fransted, J. R. Caram, E. Harel, J. Wen, R. E. Blankenship and G. S. Engel, 'Long-lived quantum coherence in photosynthetic complexes at physiological temperature', *Proceedings of the National Academy of Sciences*, vol. 107: 29 (2010), pp. 12766–70.

7   T. R. Calhoun, N. S. Ginsberg, G. S. Schlau-Cohen, Y. C. Cheng, M. Ballottari, R. Bassi and G. R. Fleming, 'Quantum coherence enabled determination of the energy landscape in light-harvesting complex II', *Journal of Physical Chemistry B*, vol. 113: 51 (2009), pp. 16291–5.

Chapter 5: Finding Nemo's home

1   Exodus 30: 34–5.
2   Quoted in A. Le Guerer, *Scent: The Mysterious and Essential Power of Smell* (New York: Kodadsha America Inc., 1994), p. 12.
3   R. Eisner, 'Richard Axel: one of the nobility in science', *P&S Columbia University College of Physicians and Surgeons*, vol. 25: 1 (2005).
4   C. S. Sell, 'On the unpredictability of odor', *Angewandte Chemie, International Edition* (English), 45: 38 (2006), pp. 6254–61.
5   K. Mori and G. M. Shepherd, 'Emerging principles of molecular signal processing by mitral/tufted cells in the olfactory bulb', *Seminars in Cell Biology*, vol. 5: 1 (1994), pp. 65–74.
6   L. Turin, *The Secret of Scent: Adventures in Perfume and the Science of Smell* (London: Faber & Faber, 2006), p. 4.
7   L. Turin, 'A spectroscopic mechanism for primary olfactory reception', *Chemical Senses*, vol. 21: 6 (1996), pp. 773–91.
8   Turin, *The Secret of Scent*, p. 176.
9   C. Burr, *The Emperor of Scent: A True Story of Perfume and Obsession* (New York: Random House, 2003).
10   A. Keller and L. B. Vosshall, 'A psychophysical test of the vibration theory of olfaction', *Nature Neuroscience*, vol. 7: 4 (2004), pp. 337–8.
11   M. I. Franco, L. Turin, A. Mershin and E. M. Skoulakis, 'Molecular vibration-sensing component in *Drosophila melanogaster* olfaction', *Proceedings of the National Academy of Science*, vol. 108: 9 (2011), pp. 3797–802.
12   J. C. Brookes, F. Hartoutsiou, A. P. Horsfield and A. M. Stoneham, 'Could humans recognize odor by phonon assisted tunneling?', *Physical Review Letters*, vol. 98: 3 (2007), p. 038101.

Chapter 6: The butterfly, the fruit fly and the quantum robin

1 F. A. Urquhart, 'Found at last: the monarch's winter home', *National Geographic*, Aug. 1976.

2 R. Stanewsky, M. Kaneko, P. Emery, B. Beretta, K. Wager-Smith, S. A. Kay, M. Rosbash and J. C. Hall, 'The cryb mutation identifies cryptochrome as a circadian photoreceptor in *Drosophila*', *Cell*, vol. 95: 5 (1998), pp. 681–92.

3 H. Zhu, I. Sauman, Q. Yuan, A. Casselman, M. Emery-Le, P. Emery and S. M. Reppert, 'Cryptochromes define a novel circadian clock mechanism in monarch butterflies that may underlie sun compass navigation', *PLOS Biology*, vol. 6: 1 (2008), e4.

4 D. M. Reppert, R. J. Gegear and C. Merlin, 'Navigational mechanisms of migrating monarch butterflies', *Trends in Neurosciences*, vol. 33: 9 (2010), pp. 399–406.

5 P. A. Guerra, R. J. Gegear and S. M. Reppert, 'A magnetic compass aids monarch butterfly migration', *Nature Communications*, vol. 5: 4164 (2014), pp. 1–8.

6 A. T. von Middendorf, *Die Isepiptesen Russlands Grundlagen zur Erforschung der Zugzeiten und Zugrichtungen der Vögel Russlands* (St Petersburg, 1853).

7 H. L. Yeagley and F. C. Whitmore, 'A preliminary study of a physical basis of bird navigation', *Journal of Applied Physics*, vol. 18: 1035 (1947).

8 M. M. Walker, C. E. Diebel, C. V. Haugh, P. M. Pankhurst, J. C. Montgomery and C. R. Green, 'Structure and function of the vertebrate magnetic sense', *Nature*, vol. 390: 6658 (1997), pp. 371–6.

9 M. Hanzlik, C. Heunemann, E. Holtkamp-Rotzler, M. Winklhofer, N. Petersen and G. Fleissner, 'Superparamagnetic magnetite in the upper beak tissue of homing pigeons', *Biometals*, vol. 13: 4 (2000), pp. 325–31.

10 C. V. Mora, M. Davison, J. M. Wild and M. M. Walker, 'Magnetoreception and its trigeminal mediation in the homing pigeon', *Nature*, vol. 432 (2004), pp. 508–11.

11 C. Treiber, M. Salzer, J. Riegler, N. Edelman, C. Sugar, M. Breuss, P. Pichler, H. Cadiou, M. Saunders, M. Lythgoe, J. Shaw and D. A. Keays, 'Clusters of iron-rich cells in the upper beak of pigeons are macrophages not magnetosensitive neurons', *Nature*, vol. 484 (2012), pp. 367–70.

12 S. T. Emlen, W. Wiltschko, N. J. Demong, R. Wiltschko and S. Bergman, 'Magnetic direction finding: evidence for its use in migratory indigo buntings', *Science*, vol. 193: 4252 (1976), pp. 505–8.

13 L. Pollack, 'That nest of wires we call the imagination: a history of some key scientists behind the bird compass sense', May 2012, p. 5: http://www.ks.uiuc.edu/History/magnetoreception.

14 Ibid., p. 6.

15 K. Schulten, H. Staerk, A. Weller, H-J. Werner and B. Nickel, 'Magnetic field dependence of the geminate recombination of radical ion pairs in polar solvents', *Zeitschrift für Physikale Chemie*, n.s., vol. 101 (1976), pp. 371–90.

16 Pollack, 'That nest of wires we call the imagination', p. 11.

17 K. Schulten, C. E. Swenberg and A. Weller, 'A biomagnetic sensory mechanism based on magnetic field modulated coherent electron spin motion', *Zeitschrift für Physikale Chemie*, n.s., vol. 111 (1978), pp. 1–5.

18 From P. Hore, 'The quantum robin', *Navigation News*, Oct. 2011.

19 N. Lambert, 'Quantum biology', *Nature Physics*, vol. 9: 10 (2013), and references therein.

20 M. J. M. Leask, 'A physicochemical mechanism for magnetic field detection by migratory birds and homing pigeons', *Nature*, vol. 267 (1977), pp. 144–5.

21 T. Ritz, S. Adem and K. Schulten, 'A model for photoreceptor-based magnetoreception in birds', *Biophysical Journal*, vol. 78: 2 (2000), pp. 707–18.

22 M. Liedvogel, K. Maeda, K. Henbest, E. Schleicher, T. Simon, C. R. Timmel, P. J. Hore and H. Mouritsen, 'Chemical magnetoreception: bird cryptochrome 1a is excited by blue light and forms long-lived radical-pairs', *PLOS One*, vol. 2: 10 (2007), e1106.

23 C. Nießner, S. Denzau, K. Stapput, M. Ahmad, L. Peichl, W. Wiltschko and R. Wiltschko, 'Magnetoreception: activated cryptochrome 1a concurs with magnetic orientation in birds', *Journal of the Royal Society Interface*, vol. 10: 88 (6 Nov. 2013), 20130638.

24 T. Ritz, P. Thalau, J. B. Phillips, R. Wiltschko and W. Wiltschko, 'Resonance effects indicate a radical-pair mechanism for avian magnetic compass', *Nature*, vol. 429 (2004), pp. 177–80.

25 S. Engels, N-L. Schneider, N. Lefeldt, C. M. Hein, M. Zapka, A. Michalik, D. Elbers, A. Kittel, P. J. Hore and H. Mouritsen, 'Anthropogenic electromagnetic noise disrupts magnetic compass orientation in a migratory bird', *Nature*, vol. 509 (2014), pp. 353–6.

26 E. M. Gauger, E. Rieper, J. J. Morton, S. C. Benjamin and V. Vedral, 'Sustained quantum coherence and entanglement in the avian compass', *Physical Review Letters*, vol. 106: 4 (2011), 040503.

27 M. Ahmad, P. Galland, T. Ritz, R. Wiltschko and W. Wiltschko, 'Magnetic intensity affects cryptochrome-dependent responses in *Arabidopsis thaliana*', *Planta*, vol. 225: 3 (2007), pp. 615–24.

28 M. Vacha, T. Puzova and M. Kvicalova, 'Radio frequency magnetic fields disrupt magnetoreception in American cockroach', *Journal of Experimental Biology*, vol. 212: 21 (2009), pp. 3473–7.

Chapter 7: Quantum genes

1   Y. M. Shtarkman, Z. A. Kocer, R. Edgar, R. S. Veerapaneni, T. D'Elia, P. F. Morris and S. O. Rogers, 'Subglacial Lake Vostok (Antarctica) accretion ice contains a diverse set of sequences from aquatic, marine and sediment-inhabiting bacteria and eukarya', *PLOS One*, vol. 8: 7 (2013), e67221.

2   J. D. Watson and F. H. C. Crick, 'Molecular structure of nucleic acids: a structure for deoxyribose nucleic acid', *Nature*, vol. 171 (1953), pp. 737–8.

3   C. Darwin, *On the Origin of Species*, ch. 4.

4   J. D. Watson and F. H. C. Crick, 'Genetic implications of the structure of deoxyribonucleic acid', *Nature*, vol. 171 (1953), pp. 964–9.

5   W. Wang, H. W. Hellinga and L. S. Beese, 'Structural evidence for the rare tautomer hypothesis of spontaneous mutagenesis', *Proceedings of the National Academy of Sciences*, vol. 108: 43 (2011), pp. 17644–8.

6   A. Datta and S. Jinks-Robertson, 'Association of increased spontaneous mutation rates with high levels of transcription in yeast', *Science*, vol. 268: 5217 (1995), pp. 1616–19.

7   J. Bachl, C. Carlson, V. Gray-Schopfer, M. Dessing and C. Olsson, 'Increased transcription levels induce higher mutation rates in a hypermutating cell line', *Journal of Immunology*, vol. 166: 8 (2001), pp. 5051–7.

8   P. Cui, F. Ding, Q. Lin, L. Zhang, A. Li, Z. Zhang, S. Hu and J. Yu, 'Distinct contributions of replication and transcription to mutation rate variation of human genomes', *Genomics, Proteomics and Bioinformatics*, vol. 10: 1 (2012), pp. 4–10.

9   J. Cairns, J. Overbaugh and S. Millar, 'The origin of mutants', *Nature*, vol. 335 (1988), pp. 142–5.

10 John Cairns on Jim Watson, Cold Spring Harbor Oral History Collection. Interview available at: http://library.cshl.edu/oralhistory/interview/james-d-watson/meeting-jim-watson/watson/.

11 J. Gribbin, *In Search of Schrödinger's Cat* (London: Wildwood House, 1984; repr. Black Swan, 2012).

12 J. McFadden and J. Al-Khalili, 'A quantum mechanical model of adaptive mutation', *Biosystems*, vol. 50: 3 (1999), pp. 203–11.

13 J. McFadden, *Quantum Evolution* (London: HarperCollins, 2000).

14 A critical review is published here: http://arxiv.org/abs/quant-ph/0101019 and our response can be found here: http://arxiv.org/abs/quant-ph/0110083.

15 H. Hendrickson, E. S. Slechta, U. Bergthorsson, D. I. Andersson and J. R. Roth, 'Amplification-mutagenesis: evidence that "directed"

adaptive mutation and general hypermutability result from growth with a selected gene amplification', *Proceedings of the National Academy of Sciences*, vol. 99: 4 (2002), pp. 2164–9.

16 e.g. J. D. Stumpf, A. R. Poteete and P. L. Foster, 'Amplification of *lac* cannot account for adaptive mutation to Lac+ in *Escherichia coli*', *Journal of Bacteriology*, vol. 189: 6 (2007), pp. 2291–9.

17 e.g. E. S. Kryachko, 'The origin of spontaneous point mutations in DNA via Löwdin mechanism of proton tunneling in DNA base pairs: cure with covalent base pairing', *International Journal Of Quantum Chemistry*, vol. 90: 2 (2002), pp. 910–23; Zhen Min Zhao, Qi Ren Zhang, Chun Yuan Gao and Yi Zhong Zhuo, 'Motion of the hydrogen bond proton in cytosine and the transition between its normal and imino states', *Physics Letters A*, vol. 359: 1 (2006), pp. 10–13.

Chapter 8: Mind

1 Interview for the *Los Angeles Times*, 14 Feb. 1995.

2 J-M. Chauvet, E. Brunel-Deschamps, C. Hillaire and J. Clottes, *Dawn of Art. The Chauvet Cave: The Oldest Known Paintings in the World* (New York: Harry N. Abrams, 1996).

3 Quoted in J. Hadamard, *Essay on the Psychology of Invention in the Mathematical Field* (Princeton: Princeton University Press, 1945). However, according to Daniel Dennett in 'Memes and the exploitation of imagination', *Journal of Aesthetics and Art Criticism*, vol. 48 (1990), pp. 127–35 (available at http://ase.tufts.edu/cogstud/dennett/papers/memeimag.htm#5), this oft-quoted passage is probably not from Mozart and is of uncertain origin. We have nevertheless decided to retain it because its author, whoever that is, managed to describe a familiar but remarkable phenomenon very nicely.

4 J. McFadden, 'The CEMI field theory gestalt information and the meaning of meaning', *Journal of Consciousness Studies*, vol. 20: 3–4 (2013), pp. 152–82.

5 Chauvet et al., *Dawn of Art*.

6 M. Kinsbourne, 'Integrated cortical field model of consciousness', in *Experimental and Theoretical Studies of Consciousness*, CIBA Foundation Symposium No. 174 (Chichester: Wiley, 2008).

7 K. Saeedi, S. Simmons, J. Z. Salvail, P. Dluhy, H. Riemann, N. V. Abrosimov, P. Becker, H.-J. Pohl, J. J. L. Morton and M. L. W. Thewalt, 'Room-temperature quantum bit storage exceeding 29 minutes using ionized donors in silicon-28', *Science*, vol. 342: 6160 (2013), ppp. 830–33.

8 D. Hofstadter, *Gödel, Escher, Bach: An Eternal Golden Braid* (New York: Basic Books, 1999; first publ. 1979).

9   R. Penrose, *Shadows of the Mind: A Search for the Missing Science of Consciousness* (Oxford: Oxford University Press, 1994).

10   S. Hameroff, 'Quantum computation in brain microtubules? The Penrose–Hameroff "Orch OR" model of consciousness', *Philosophical Transactions of the Royal Society Series A*, vol. 356: 1743 (1998), pp. 1869–95; S. Hameroff and R. Penrose, 'Consciousness in the universe: a review of the "Orch OR" theory', *Physics of Life Reviews*, vol. 11 (2014), pp. 39–78.

11   M. Tegmark, 'Importance of quantum decoherence in brain processes', *Physical Review E*, vol. 61 (2000), pp. 4194–206.

12   See e.g. A. Litt, C. Eliasmith, F. W. Kroon, S. Weinstein and P. Thagard, 'Is the brain a quantum computer?', *Cognitive Science*, vol. 30: 3 (2006), pp. 593–603.

13   G. Bernroider and J. Summhammer, 'Can quantum entanglement between ion transition states effect action potential initiation?', *Cognitive Computation*, vol. 4 (2012), pp. 29–37.

14   McFadden, *Quantum Evolution*; J. McFadden, 'Synchronous firing and its influence on the brain's electromagnetic field: evidence for an electromagnetic theory of consciousness', *Journal of Consciousness Studies*, vol. 9 (2002), pp. 23–50; S. Pockett, *The Nature of Consciousness: A Hypothesis* (Lincoln, NE: Writers Club Press, 2000); E. R. John, 'A field theory of consciousness', *Consciousness and Cognition*, vol. 10: 2 (2001), pp. 184–213; J. McFadden, 'The CEMI field theory closing the loop', *Journal of Consciousness Studies*, vol. 20: 1–2 (2013), pp. 153–68.

15   McFadden, 'The CEMI field theory gestalt information and the meaning of meaning'.

16   C. A. Anastassiou, R. Perin, H. Markram and C. Koch, 'Ephaptic coupling of cortical neurons', *Nature Neuroscience*, vol. 14: 2 (2011), pp. 217–23; F. Frohlich and D. A. McCormick, 'Endogenous electric fields may guide neocortical network activity', *Neuron*, vol. 67: 1 (2010), pp. 129–43.

17   McFadden, 'The CEMI field theory closing the loop'.

18   W. Singer, 'Consciousness and the structure of neuronal representations', *Philosophical Transactions of the Royal Society B: Biological Sciences*, vol. 353: 1377 (1998), pp. 1829–40.

Chapter 9: How life began

1   S. L. Miller, 'A production of amino acids under possible primitive earth conditions', *Science*, vol. 117: 3046 (1953), pp. 528–9.

2   G. Cairns-Smith, *Seven Clues to the Origin of Life: A Scientific Detective Story* (Cambridge: Cambridge University Press, 1985; new edn 1990).

3   McFadden, *Quantum Evolution*; J. McFadden and J. Al-Khalili, 'Quantum coherence and the search for the first replicator', in D. Abbott, P. C. Davies and A. K. Patki, eds, *Quantum Aspects of Life* (London: Imperial College Press, 2008).
4   A. Patel, 'Quantum algorithms and the genetic code', *Pramana Journal of Physics*, vol. 56 (2001), pp. 367–81; available at http://arxiv.org/pdf/quant-ph/0002037.pdf.

Chapter 10: Quantum biology: life on the edge of a storm

1   M. B. Plenio and S. F. Huelga, 'Dephasing-assisted transport: quantum networks and biomolecules', *New Journal of Physics*, vol. 10 (2008), 113019; F. Caruso, A. W. Chin, A. Datta, S. F. Huelga and M. B. Plenio, 'Highly efficient energy excitation transfer in light-harvesting complexes: the fundamental role of noise-assisted transport', *Journal of Chemical Physics*, vol. 131 (2009), 105106–21.
2   M. Mohseni, P. Rebentrost, S. Lloyd and A. Aspuru-Guzik, 'Environment-assisted quantum walks in photosynthetic energy transfer', *Journal of Chemical Physics*, vol. 129: 17 (2008), 174106.
3   B. Misra and G. Sudarshan, 'The Zeno paradox in quantum theory', *Journal of Mathematical Physics*, vol. 18 (1977), p. 746: http://dx.doi.org/10.1063/1.523304.
4   S. Lloyd, M. Mohseni, A. Shabani and H. Rabitz, 'The quantum Goldilocks effect: on the convergence of timescales in quantum transport', arXiv preprint, arXiv:1111.4982, 2011.
5   A. W. Chin, S. F. Huelga and M. B. Plenio, 'Coherence and decoherence in biological systems: principles of noise-assisted transport and the origin of long-lived coherences', *Philosophical Transactions of the Royal Society A*, vol. 370 (2012), pp. 3658–71; A. W. Chin, J. Prior, R. Rosenbach, F. Caycedo-Soler, S. F. Huelga and M. B. Plenio, 'The role of non-equilibrium vibrational structures in electronic coherence and recoherence in pigment-protein complexes', *Nature Physics*, vol. 9: 2 (2013), pp. 113–18.
6   E. J. O'Reilly and A. Olaya-Castro, 'Non-classicality of molecular vibrations activating electronic dynamics at room temperature', *Nature Communications*, vol. 5 (2014), article no. 3012.
7   I. Stewart, *Does God Play Dice?: The New Mathematics of Chaos* (Harmondsworth: Penguin UK, 1997); S. Kauffman. *The Origins of Order: Self-Organization and Selection in Evolution* (New York: Oxford University Press, 1993); J. Gleick, *Chaos: Making a New Science* (New York: Random House, 1997).
8   M. O. Scully, K. R. Chapin, K. E. Dorfman, M. B. Kim and A. Svidzinsky, 'Quantum heat engine power can be increased by noise-induced

coherence', *Proceedings of the National Academy of Sciences*, vol. 108: 37 (2011), pp. 15097–100.

9   K. E. Dorfman, D. V. Voronine, S. Mukamel and M. O. Scully, 'Photosynthetic reaction center as a quantum heat engine', *Proceedings of the National Academy of Sciences*, vol. 110: 8 (2013), pp. 2746–51.

10 M. Ferretti, V. I. Novoderezhkin, E. Romero, R. Augulis, A. Pandit, D. Zigmantas and R. Van Grondelle, 'The nature of coherences in the B820 bacteriochlorophyll dimer revealed by two-dimensional electronic spectroscopy', *Physical Chemistry Chemical Physics*, vol. 16 (2014), pp. 9930–9.

11 C. R. Pudney, A. Guerriero, N. J. Baxter, L. O. Johannissen, J. P. Waltho, S. Hay and N. S. Scrutton, 'Fast protein motions are coupled to enzyme H-transfer reactions', *Journal of the American Chemical Society*, vol. 135 (2013), pp. 2512–17.

12 J. P. Klinman and A. Kohen, 'Hydrogen tunnelling links protein dynamics to enzyme catalysis', *Annual Review of Biochemistry*, vol. 82 (2013), pp. 471–96.

13 R. Armstrong and N. Spiller, 'Living quarters', *Nature*, vol. 467 (2010), pp. 916–19.

14 S. Ludec, *The Mechanism of Life* (London: William Heinemann, 1914).

15 T. Toyota, N. Maru, M. M. Hanczyc, T. Ikegami and T. Sugawara, 'Self-propelled oil droplets consuming "fuel" surfactant', *Journal of the American Chemical Society*, vol. 131: 14 (2009), pp. 5012–13.

16 I. A. Chen, K. Salehi-Ashtiani and J. W. Szostak, 'RNA catalysis in model protocell vesicles', *Journal of the American Chemical Society*, vol. 127: 38 (2005), pp. 13213–19.

17 R. J. Peters, M. Marguet, S. Marais, M. W. Fraaije, J. C. van Hest and S. Lecommandoux, 'Cascade reactions in multicompartmentalized polymersomes'. *Angewandte Chemie International Edition* (English), vol. 53: 1 (2014), pp. 146–50.

18 D. Hayes, G. B. Griffin and G. S. Engel, 'Engineering coherence among excited states in synthetic heterodimer systems', *Science*, vol. 340: 6139 (2013), pp. 1431–4.

19 Dorfman et al., 'Photosynthetic reaction center as a quantum heat engine'.

20 C. Creatore, M. A. Parker, S. Emmott and A. W. Chin, 'An efficient biologically-inspired photocell enhanced by quantum coherence', arXiv preprint, arXiv:1307.5093, 2013.

21 C. Tan, S. Saurabh, M. P. Bruchez, R. Schwartz and P. Leduc, 'Molecular crowding shapes gene expression in synthetic cellular nanosystems', *Nature Nanotechnology*, vol. 8: 8 (2013), pp. 602–8; M. S. Cheung, D. Klimov and D. Thirumalai, 'Molecular crowding enhances native state stability and refolding rates of globular proteins', *Proceedings of the National Academy of Sciences*, vol. 102: 13 (2005), pp. 4753–8.

Index

Page numbers in *italics* refer to illustrations.

INDEX

Tetrahymena, 283
thermodynamics: entropy, 34; free
energy, 34n; in living cells, 40; life
as, 35, 37, *314*, *316*; principles, 33;
Schrödinger on, 53, 55, 58, 296;
science of, 32–3, 306
thought, mechanics of, 248–54
3D printers, 287–8
thylakoids, 124
Tinba, 288–9
transcription, 225, 227
transition state theory (TST), 71–2,
75–7, 83–4
translation, 225
tritium, 97
trout, 178
tryptophan, 196
tubulin proteins, 263
tunnelling, 90–3, *92*: atoms, 94; DNA
mutation rates, 223–4, 232–3;
elastic, 160; electrons, 93–4, 95,
98, 160–1, 167, 223, 327; enzyme
activity, 95, 98, 99, 135, 160, 163,
223, 232, 311, 331; in biology, 20,
23, 93–4, 99–101; inelastic, 160–1,
168; kinetic isotope effect, 96,
98, 232; process, 10–11, 121, 235;
proto-enzyme, 290–1; protons, 94,
95–6, 98, 223, 232, 233, 235, 311,
312, 327; proto-self-replicator, 293
Turin, Luca, 159–62, 163–6, 244
turtles, 3, 178
two-slit experiment, 105–18, *107*,
110, *111*, 302, 309

Uncertainty Principle, Heisenberg's,
17n, 46–7, 52
Urey, Harold, 278–9
Urquhart, Fred and Norah, 170–4

Vedral, Vlatko, *19*, 199
Venter, Craig, 41–2, 281,
319–20, 328
Vikings, 271–2
Virchow, Rudolf, 36
vital activities, 66
vital force, 65, 69, 103
vitalism, 36–7, 50, 103, 315
voles, 142
Vosshall, Leslie, 164, 166
Vostok, Lake, 203–5, 207,
214, 235–6
Vostok Antarctic station, 202–3
Voyager missions, 24–5, 204

Wallace, Alfred Russel, 215n
Watson, James, 39, 208, 219, 226
wave function, 47–50, 51n,
115–18, 121
wave mechanics, 47
wave–particle duality, 8–9, 10,
105, 115
wave properties, 8–11, 45, 91–4,
98, 105
What Is Life? (Schrödinger), 53, 56,
59, 213
*What Is Life? The Next Fifty
Years*, 57
Wiltschko, Wolfgang and Roswitha,
5, 6, 14, 18, 19, 180–3, 194–8
Wöhler, Friedrich, 37
Wright, Robert H., 156–7, 167

Yeagley, Henry, 178
yeast, 64, 65–6, 69, 95, 225
Young, Thomas, 106

Zeno's paradox, 301–2
zombies, 242

Professor Jim Al-Khalili, OBE is an academic, author and broadcaster. He is a leading theoretical physicist based at the University of Surrey, where he teaches and carries out research in quantum mechanics. He has written a number of popular science books, including *Pathfinders: The Golden Age of Arabic Science, Quantum: A Guide for the Perplexed* and *Paradox: The Nine Greatest Enigmas in Science*. He has presented several television and radio documentaries, including the BAFTA-nominated *Chemistry: A Volatile History* and *The Secret Life of Chaos*.

Professor Johnjoe McFadden is Professor of Molecular Genetics at the University of Surrey and is the editor of leading textbooks on both molecular biology and systems biology of tuberculosis. For over a decade, he has specialized in examining tuberculosis and meningitis, inventing the first successful molecular test for the latter and winning the Royal Society Wolfson Research Merit Award for his findings. He is the author of *Quantum Evolution* and co-editor of *Human Nature: Fact and Fiction*, and writes for the *Guardian* on topics including GM crops, psychedelic drugs and quantum mechanics.